Lecture Notes in Mathematics

Edited by A. Dold and B. Eckmann

T0216203

1398

A.C. Kim   B.H. Neumann   (Eds.)

Groups – Korea 1988

Proceedings of a Conference on Group Theory,
held in Pusan, Korea, August 15–21, 1988

Springer-Verlag

Berlin Heidelberg New York London Paris Tokyo Hong Kong

Editors

Ann Chi Kim
Department of Mathematics, Pusan National University
Pusan 607, Republic of Korea

Bernhard H. Neumann
Mathematics Research Section, School of Mathematical Sciences
Australien National University
Canberra, ACT 2601, Australia

Mathematics Subject Classification (1980): 20EXX, 20FXX, 20DXX, 20BXX, 20NXX, 05C25

ISBN 3-540-51695-6 Springer-Verlag Berlin Heidelberg New York
ISBN 0-387-51695-6 Springer-Verlag New York Berlin Heidelberg

© Springer-Verlag Berlin Heidelberg 1989
Printed in Germany

Printing and binding: Druckhaus Beltz, Hemsbach/Bergstr.
2146/3140-543210 – Printed on acid-free paper

PREFACE

The first international conference on the theory of groups held in Korea, "Groups – Korea 1983", was reported on in the Springer–Verlag Lecture Notes in Mathematics, Volume 1098. "Groups – Korea 1988" is the second such conference. It was held in the Commodore Hotel, Pusan, in August 1988. There were 16 invited one–hour lectures and 17 contributed half-hour seminar–type talks. They are listed in Appendix A. The participants are listed in Appendix B.

"Groups – Korea 1988" was financially supported by the Korean Educational Ministry, the Korea Science and Engineering Foundation, the Korean Mathematical Society, the Pusan Chamber of Commerce & Industry, and the Pusan National University. We record our thanks to these institutions, and to their officers, to whose support the success of the conference owes much. We are also grateful to Dr Jost–Gert Glombitza, of the Deutsche Forschungsgemeinschaft, and to the Deutsche Forschungsgemeinschaft itself, for supporting the five invited speakers from the Federal Republic of Germany. Our thanks also go to the President of the Pusan Chamber of Commerce & Industry, Mr Chung Whan Choi, for special financial contributions, and to Dr Jung Rae Cho, a member of the organizing committee, who typed all manuscripts for the proceedings with great care. Finally, we thank all the staff of the Commodore Hotel, who proved very helpful before and after the conference, and showed wonderful hospitality during the conference itself.

A.C. Kim
B.H. Neumann
Editors

TABLE OF CONTENTS

PROCEEDINGS OF 'GROUPS – KOREA 1988'

PUSAN, August 1988

INFINITE FACTORIZED GROUPS

Bernhard Amberg

A group G is called *factorized* if it can be written as the product of two of its subgroups A and B:

$$G = AB = \{ab \mid a \in A \text{ and } b \in B\}.$$

In the past thirty-five years an increasing number of facts about factorized groups have been discovered particularly for infinite groups. The following survey is a slightly extended version of [14] (see also Robinson [71] and N.S. Chernikov [48]). The notation is standard and may for instance be found in [68] and [70].

I. Existence of Factorizations.

Every group G has at least one factorization $G = AB$, where one of the two subgroups A and B is trivial and the other is the whole group. It is also clear that every infinite group which has all its proper subgroups finite, cannot have any other non-trivial factorization. Thus, in particular, the Prüfer groups of type p^∞ and also the so-called Tarski p-group have only the trivial factorization (A *Prüfer group of type* p^∞ for the prime p is a group which is isomorphic with the multiplicative group of complex p^n-th roots of unity for $n = 0, 1, 2, 3, \cdots$; all proper subgroups of such a group are cyclic of order p^n for some n and they form a chain. A *Tarski p-group* is an infinite group which has all its proper subgroups cyclic of order p for a very large prime p; the existence of such has been established by Ol'shanskij and Rips).

It is well-known that an abelian group which has no non-trivial factorization must be a cyclic p-group or a Prüfer group of type p^∞ for some prime p (see [75], 13.1.6). Using this it is not difficult to see the following:

Every uncountable soluble group and every infinite nilpotent group which is not a Prüfer group of type p^∞ always has at least one non-trivial factorization.

In [67] Rédei considers the structure of groups which are the product of two cyclic subgroups. The groups which are the product of two locally cyclic, torsion-free subgroups are described in Sysak [82].

For some simple groups all possible factorizations are known. For instance in [83] Thiel has found all the factorizations of the groups $PSL(2,K)$, $PSL(3,K)$ and $PSU(3,K)$, where K is a locally finite field. For finite K this is also contained in Itô [58] and Blaum [28]. There are many more papers on the factorizations of the various finite simple groups.

II. The Main Problem.

The main problem about factorized groups is the following question:

Let $G = AB$ be a given factorized group and suppose that the structure of the two subgroups A and B is known. What can then be said about the structure of the factorized group $G = AB$? If, for instance, A and B have a certain group theoretical property $\mathfrak{X}$, when does it follow that G has a certain group theoretical property $\mathfrak{Y}$?

Statements of this type can be trivial: for instance, if A and B are finite (finitely generated, countable), then $G = AB$ is finite (finitely generated, countable). For many other group theoretical properties $\mathfrak{X}$ almost nothing can be said. In particular it is difficult to find examples and counterexamples. In the following we consider some group theoretical properties $\mathfrak{X}$ for which some interesting theorems can be proved.

First we remark that the above problems usually become trivial when one of the two subgroups A and B is subnormal in $G = AB$. This can be seen from the following simple lemma.

Lemma. *Let the group theoretical property $\mathfrak{X}$ be inherited by subgroups, epimorphic images and extensions. If the group $G = AB$ is the product of two $\mathfrak{X}$-subgroups A and B, one of which is subnormal in G, then G is an $\mathfrak{X}$-group.*

Proof. Suppose that the subgroup A is subnormal in G and induct on the length of some subnormal series from A to G. If A is normal in G, then

$$G/A = AB/A \cong B/(A \cap B)$$

is an $\mathfrak{X}$-group. Since also A is an $\mathfrak{X}$-group, G is an $\mathfrak{X}$-group.

If A is not normal in G, then A is normal in some subnormal subgroup S of G which has a subnormal series in G with length less than that of A. Then $G = SB$ and since $B \cap S$ is an $\mathfrak{X}$-group, also

$$(S \cap B)/(A \cap B) = (S \cap B)/(A \cap B \cap S) \cong (S \cap B)A/A = (S \cap AB)/A = S/A$$

is an $\mathfrak{X}$-group. Since A is an $\mathfrak{X}$-group, also S is an $\mathfrak{X}$-group. By induction it follows that $G = SB$ is an $\mathfrak{X}$-group. This proves the lemma.

The most satisfying result about products of groups is the following famous theorem.

(1) (Itô [57]). *If $G = AB$ where A and B are abelian subgroups of the group G, then G is metabelian.*

This is proved by a surprisingly short commutator calculation (see also [75]. 13.3.2, or [56], p.674). This calculation so far could not be extended to products of nilpotent groups (of class 2).

However, in a slightly more general situation, the (almost) solubility of the group $G = AB$ can still be obtained. A group G is *almost soluble (almost central)* if it contains a soluble (central) subgroup of finite index.

(2) (Chernikov [35]). *If $G = AB$ where A and B are almost central subgroups of the group G, then G is almost soluble.*

The proof is by induction on the sum of the finite indices $|A/Z(A)|$ and $|B/Z(B)|$ of the centers $Z(A)$ and $Z(B)$ of A and B; the beginning of the induction is Itô's theorem. It can also be shown that the solubility length of the soluble subgroup of G of finite index is bounded by a function of $|A/Z(A)|$ and $|B/Z(B)|$. However it remains open whether the group $G = AB$ is in fact almost metabelian. – In this connection Wilson has shown in [92] that every residually finite product of two almost abelian subgroups is almost metabelian. (A group G is *residually finite* if the intersection $J(G)$ of all subgroups of finite index in G is trivial).

There is another non-simplicity criterion which holds for arbitrary factorized groups. Recall that a group is an *FC-group* if all its elements have only finitely many conjugates in the group. A group G is *hyperabelian* if it has an ascending invariant series leading from 1 to G with abelian factors.

(3) (Zajtsev [97] and [99] and Chernikov [47]). *Let the infinite group $G = AB$ be the product of an abelian subgroup A and an FC-subgroup B with non-trivial center $Z(B)$. Then the normal closure $C = Z(B)^G$ of $Z(B)$ has an ascending G-invariant series with abelian factors; in particular $AC = A(B \cap AC)$ is hyperabelian and G has a non-trivial abelian normal subgroup.*

The most famous theorem on products of finite groups is the following.

(4) (Wielandt [87] and Kegel [60]). *If $G = AB$ where A and B are finite nilpotent subgroups of the group G, then G is soluble.*

The proof makes use of many techniques of finite group theory (see also [75], 13.2.9, and [56], p.674). Some extensions of the Kegel-Wielandt theorem to certain classes of locally finite groups can for instance be found in Kegel [61], Amberg [7] and Chernikov [34]. However, it seems to be unknown at present whether the theorem of Kegel and Wielandt holds for arbitrary infinite groups.

Using the classification of finite simple groups it was possible to prove a famous conjecture of Szép (see also Kazarin [59]).

(5) (Fisman and Arad [50]). *If $G = AB$ where A and B are finite subgroups with non-trivial centers $Z(A)$ and $Z(B)$, then G is not simple.*

In the situation of the Kegel-Wielandt theorem it is not known how the solubility length of the (finite) soluble group $G = AB$ depends on the nilpotency classes α and β of the nilpotent subgroups A and B. Is it perhaps $\alpha + \beta$? Some indications in this direction are contained in the following result (for an extension to certain locally finite groups see Amberg [7]).

(6) (Pennington [65]). *If $G = AB$ where A and B are finite nilpotent subgroups of the group G with nilpotency classes α and β, then the $(\alpha + \beta)$-th term $G^{(\alpha+\beta)}$ of the derived series of G is a π-group where $\pi = \pi A \cap \pi B$.*

Here, as usual πY denotes the set of primes p for which there exists an element of order p in the group Y.

In view of all the above results one is tempted to ask the following question.

Problem. Let the group $G = AB$ be the product of the subgroups A and B which contain nilpotent normal subgroups A_1 of A and B_1 of B with classes α and β and finite indices $|A : A_1|$ and $|B : B_1|$. Does then G have a soluble normal subgroup of finite index with derived length a function of α and β?

III. Periodic Groups.

A group is *periodic* if all its elements have finite order, it is *locally finite* if its finitely generated subgroups are finite. The product of two periodic (locally finite) groups need not be periodic (locally finite), as the following example shows.

(7) (Suchkov [79] and [80]). *There exists a countable group $G = AB$ where A and B are locally finite subgroups of G, but G is not periodic. In fact, G contains every countable free group and also 2-subgroups which are not locally finite.*

Construction. The construction is as follows. Let $\mathbf{Z}$ be the set of integers. For each k in $\mathbf{Z}$ and $n = 0, 1, 2, \cdots$ we define the following subsets of $\mathbf{Z}$:

$$U_n^{(k)} = \{z \mid z \in \mathbf{Z},\ 3^n(2k - 1) + 1 \leq z \leq 3^n(2k + 1)\}$$
$$V_n^{(k)} = \{z \mid z \in \mathbf{Z},\ 3^n \cdot 2k + 1 \leq z \leq 3^n \cdot 2(k + 1)\}.$$

It is easy to see that each of these sets contains $2 \cdot 3^n$ elements and that for each n the $U_n^{(k)}$ for various k and also the $V_n^{(k)}$ for various k form partitions of $\mathbf{Z}$. It is also clear that for $n > 0$

$$U_n^{(k)} = U_{n-1}^{(3k-1)} \cup U_{n-1}^{(3k)} \cup U_{n-1}^{(3k+1)} \qquad (*)$$
$$V_n^{(k)} = V_{n-1}^{(3k)} \cup V_{n-1}^{(3k+1)} \cup V_{n-1}^{(3k+2)}.$$

The following are subgroups of the symmetric group $S(\mathbf{Z})$ on $\mathbf{Z}$:

$$A_n = \{g \mid g \in S(\mathbf{Z}),\ U_n^{(k)} g = U_n^{(k)},\ k \in \mathbf{Z}\}$$
$$B_n = \{g \mid g \in S(\mathbf{Z}),\ V_n^{(k)} g = V_n^{(k)},\ k \in \mathbf{Z}\}$$

Then A_n and B_n are for each n isomorphic with certain cartesian products of symmetric groups $S_{2 \cdot 3^n}$ of degree $2 \cdot 3^n$. By the above relations $(*)$ for each $n > 0$ we have $A_{n-1} \subset A_n$ and $B_{n-1} \subset B_n$.

It follows that

$$A = \bigcup_{n=0}^{\infty} A_n \quad \text{and} \quad B = \bigcup_{n=0}^{\infty} B_n$$

are locally finite subgroups of $S(\mathbf{Z})$. Now it is not difficult to prove the following

Lemma. $A_{n-1}B_{n-1} \subset B_n A_n$ and $B_{n-1}A_{n-1} \subset A_n B_n$ for $n > 0$.

The lemma implies that $AB = BA = G$ is a group. To see that it contains elements of infinite order we consider

$$U_0^{(k)} = \{2k, 2k+1\} \quad \text{and} \quad V_0^{(k)} = \{2k+1, 2k+2\}.$$

Define the element $a_0 \in A_0$ by

$$(2k)a_0 = 2k+1 \quad \text{and} \quad (2k+1)a_0 = 2k \quad \text{for each } k \in \mathbf{Z},$$

and define the element $b_0 \in B_0$ by

$$(2k+1)b_0 = 2k+2 \quad \text{and} \quad (2k+2)b_0 = 2k+1.$$

Then

$$(2k)a_0 b_0 = 2k+2 \quad \text{for each } k \in \mathbf{Z},$$

so that the element $a_0 b_0 \in AB = G$ has infinite order.

This construction will now be slightly modified to produce an example $G^* = A^* B^*$ which is countable. Since A and B are locally finite, we can find finite subgroups A_0^* of A and B_0^* of B such that $a_0 \in A_0^*$ and $b_0 \in B_0^*$. Write

$$A_0^* B_0^* = \{b_1 a_1, \cdots, b_k a_k \mid a_i \in A, \; b_i \in B \text{ for } i = 1, \cdots, k\}.$$

Since A and B are locally finite, the subgroups

$$A_1^* = \langle A_0^*, a_1, \cdots, a_k \rangle \quad \text{and} \quad B_1^* = \langle B_0^*, b_1, \cdots, b_k \rangle$$

are finite. Clearly

$$A_0^* \subset A_1^*, \quad B_0^* \subset B_1^*, \quad A_0^* B_0^* \subset B_1^* A_1^*.$$

In this way we construct an ascending series of finite subgroups with

$$A_0^* \subset A_1^* \subset A_2^* \subset A_3^* \subset \cdots \quad \text{and} \quad B_0^* \subset B_1^* \subset B_2^* \subset B_3^* \subset \cdots$$

such that

$$A_0^* B_0^* \subset B_1^* A_1^* \subset A_2^* B_2^* \subset B_3^* A_3^* \subset \cdots.$$

Then

$$A^* = \bigcup_{i=0}^{\infty} A_i^* \quad \text{and} \quad B^* = \bigcup_{i=0}^{\infty} B_i^*$$

are locally finite subgroups, and $A^* B^* = B^* A^* = G^*$ is a countable group with the desired properties, since it contains the element $a_0 b_0$ of infinite order.

It is shown in [80] that G^* even contains every countable free group and 2-subgroups which are not locally finite.

On the other hand soluble products of two periodic groups are always periodic. This follows from the following theorem. A group G is a π-*group* if the orders of all its elements contain only primes from the set of primes π.

(8) (Sysak [81]) *If $G = AB$ is a hyperabelian group where A and B are π-subgroups of G for a set of primes π, then G is a π-group.*

This result has an interesting consequence.

Corollary. *If $G = AB$ is a soluble group where A and B are subgroups of finite exponent, then G has finite exponent.*

Proof. Let G^*, A^* and B^* be cartesian products of countably many copies of G, A and B respectively. Then $G^* = A^*B^*$ is soluble and the subgroups A^* and B^* have finite exponent. By Theorem 8 G^* is periodic. But this implies that the group G must have finite exponent.

Some further results on products of abelian subgroups of finite exponent can be found in Brisley and MacDonald [29], Holt and Howlett [54] and Howlett [55].

In [81] Sysak constructs a locally soluble group $G = AB$ where A and B are p'-groups for the complementary set p' of the prime p, but G is not a p'-group. The following questions remain open.

Problems. (a) Is every locally soluble product of two periodic subgroups periodic? Is every product of two soluble and periodic subgroups periodic?

(b) Let $\mathfrak{X}$ be a class of groups which is closed under the forming of subgroups, epimorphic images and extensions. Is every soluble product $G = AB$ of two $\mathfrak{X}$-subgroups A and B likewise an $\mathfrak{X}$-group? (Note that Suchkov's example shows that for non-soluble groups $G = AB$ and the class $\mathfrak{X}$ of periodic groups the answer to the last question is negative).

IV. Minimum and Maximum Conditions.

The following lemma is elementary (see Amberg [2] or [3]).

Lemma. *Let the group $G = AB$ be the product of its subgroups A and B.*

(a) *If A and B satisfy the maximum condition on subgroups, then G satisfies the maximum condition on normal subgroups.*

(b) *If A and B satisfy the minimum condition on subgroups, then G satisfies the minimum condition on normal subgroups.*

Proof. (of (a), the proof of (b) is similar). If N_i is an ascending chain of normal subgroups of G, then

$$A \cap N_i = A \cap N_{i+1} \text{ and } B \cap AN_i = B \cap AN_{i+1} \text{ for almost all } i.$$

By the modular law

$$AN_i = AN_i \cap AB = A(B \cap AN_i) = A(B \cap AN_{i+1}) = AN_{i+1}$$

for almost all i. Therefore

$$N_i = N_i(A \cap N_i) = N_i(A \cap N_{i+1}) = N_{i+1} \cap AN_i = AN_{i+1} \cap N_{i+1} = N_{i+1}$$

for almost all i. This proves the lemma.

The lemma raises the question whether every product of two subgroups $G = AB$ with maximum (minimum) condition on subgroups always has the maximum (minimum) condition on subgroups. This question is open even when A and B are soluble. However it has a positive answer when the whole group G is soluble. We first consider the minimum condition.

A soluble group G satisfies the minimum condition on subgroups if and only if it is a *Chernikov group* [*], i.e. there exists a normal subgroup $J(G)$ of G such that the factor group $G/J(G)$ is finite and $J(G)$ is the direct product of finitely many Prüfer groups of type p^∞ for finitely many primes p. (In any group G denote by $J(G)$ the intersection of all subgroups of finite index in G and call this the *finite residual* of G).

In [76] Sesekin has shown that the product $G = AB$ of two abelian subgroups A and B with minimum condition is a (metabelian) Chernikov group. In Amberg [2] and [3] it is shown that every hyper-(almost-abelian) product $G = AB$ of two Chernikov groups is a Chernikov group and that $J(G) = J(A)J(B)$; for soluble groups G this was proved by Kegel already around 1965 but not published. The most general result on products of Chernikov groups obtained so far concerns so-called locally graduated groups. A group G is *locally graduated* if every non-trivial finitely generated subgroup of G has a non-trivial finite epimorphic image. The class of these groups is very large and contains the locally finite, the locally soluble, the residually finite and the linear groups.

(9) (Chernikov [32]). *If the locally graduated group $G = AB$ is the product of two Chernikov subgroups A and B, then G is a Chernikov group, so that*

$$J(G) = J(A)J(B).$$

A similar theorem holds for products of periodic groups with minimum condition on p-subgroups for every prime p. A soluble group G is periodic with minimum condition on p-subgroups for every prime p if and only if the finite residual $J(G)$ is the direct product of Prüfer groups of type p^∞, for each prime p only finitely many (but possibly infinitely many primes), and $G/J(G)$ has finite maximal p-subgroups for each prime p (see [63], 3.18). In particular we note

---

[*]named after S.N.Chernikov

(10) (Amberg [8] and [9], Chernikov [38] and [48]). *If the soluble group $G = AB$ is the product of two periodic subgroups A and B with minimum condition on p-subgroups for every prime p, then G is periodic with minimum condition on p-subgroups for every prime p and*

$$J(G) = J(A)J(B).$$

We turn now to the maximum condition. A soluble group G satisfies the maximum condition on subgroups if and only if it is *polycyclic*, i.e. there exists a finite series G_i of G

$$1 = G_0 \triangleleft G_1 \triangleleft G_2 \triangleleft \cdots \triangleleft G_n = G$$

such that the factors G_i/G_{i-1} are cyclic. The number of infinite cyclic factors G_i/G_{i-1} in any such series is an invariant of G, called its *torsion-free rank* or *Hirsch-number* $r_0(G)$. – A group G is *almost polycyclic* if it contains a polycyclic normal subgroup N of finite index in G. In this case put $r_o(G) = r_o(N)$.

It was shown around 1972 by Amberg and independently by Sesekin that every product of two finitely generated abelian groups is (metabelian and) polycyclic (see [77], [2] and [3]). In the last two papers it is even shown that every soluble product $G = AB$ of two polycyclic subgroups A and B, one of which is nilpotent, is always polycyclic. Finally the condition that A or B is nilpotent, was removed in [64] and [95]. The proof in [64] even holds for an almost soluble group G.

(11) (Lennox and Roseblade [64], see also Zajtsev [95]). *If the almost soluble group $G = AB$ is the product of two (almost) polycyclic subgroups A and B, then G is (almost) polycyclic.*

(11+) (Amberg [2], [4] and [12]). *If $G = AB$ is almost polycyclic, then the torsion-free rank $r_0(G)$ of G satisfies*

$$r_0(G) = r_0(A) + r_0(B) - r_0(A \cap B).$$

Comments on the proof of Theorem 11.

In order to explain the main reduction arguments that are used in proving theorems about (almost) soluble factorized groups we discuss briefly the proof of the following special case of Theorem 11:

(*) *If $G = AB$ is a soluble group where A and B are polycyclic and A or B is nilpotent, then G is polycyclic* (Amberg [2] and [3]).

Assume that the group G in (*) is not polycyclic. Let $G = AB$ be a counterexample with minimal derived length, and let $K \neq 1$ be the last non-trivial term of the derived series of G. Then $G/K = (AK/K)(BK/K)$ is polycyclic, so that the abelian normal subgroup K of G cannot be finitely generated.

In this situation it is convenient to consider the so-called *factorizer* of K in G; this is the subgroup

$$X = X(K) = AK \cap BK$$

It is not difficult to see that it has the '*triple factorization*'

$$X = K(A \cap BK) = K(B \cap AK) = (A \cap BK)(B \cap AK).$$

Since K is not finitely generated, X is also a counterexample. Therefore we may suppose without loss of generality that

$$(\alpha) \qquad\qquad\qquad G = AK = BK = AB$$

Clearly $C = (A \cap K)(B \cap K)$ is normal in $AK = BK = G$. Since $A \cap K$ and $B \cap K$ are finitely generated, C is abelian and finitely generated. Therefore also $G/C = (AC/C)(BC/C)$ is not polycyclic. Since this group is also a counter example with a triple factorization of type (α) we may in addition suppose that

$$(\beta) \qquad\qquad\qquad A \cap K = B \cap K = 1$$

Therefore A and B are complements of the normal subgroup K of G. In particular A and B are isomorphic and therefore both nilpotent.

In this situation one usually tries to apply cohomological arguments to show that A and B are conjugate, which immediately implies $A = B = G$, a contradiction. For this, in particular some 'vanishing theorems' of D. Robinson play a decisive role (see [69] and [73]). These guarantee the vanishing of the first cohomology group $H^1(A, K)$ under certain conditions. This is for example the case when A is nilpotent, K is a noetherian A-module and $K = [K, A]$ (see [69 or [73]).

In the original proof of (*) a contradiction was obtained in the following way. By an earlier result of Kegel [62] a finite triply factorized group of type (α) with three nilpotent factors A, B and K is likewise nilpotent; see also Theorem 21 below. Thus every finite epimorphic image of the counterexample $G = AB$ is nilpotent. Since A and B are polycyclic, $G = AB$ is a finitely generated soluble group. Now a well-known result of Robinson and Wehrfritz yields that G is nilpotent and hence polycyclic (see [70] p.459). This contradiction proves (*).

In the proof of the general statement of Theorem 11 one needs much deeper results about finitely generated modules over polycyclic groups, in particular the theorem of Roseblade in [74] that a simple module over a polycyclic group is finite. Thus the Theorem of Lennox-Roseblade-Zajtsev cannot be regarded as an elementary result.

The following problems remain open.

Problems. (a) Is every product of two polycyclic groups always polycyclic?
(b) Is every product of two Chernikov groups always a Chernikov group?

V. Nilpotent-by-Polycyclic Groups.

A group G is called *nilpotent-by-polycyclic* if it contains a nilpotent normal subgroup with poly-cyclic factor group. In such a group G also the Fitting subgroup Fit G is a nilpotent characteristic subgroup of G with polycyclic factor group $G/$Fit G (Fit G denotes the product of all nilpotent nor-mal subgroups of G). It follows that an extension of a nilpotent-by-polycyclic group by a polycyclic group is likewise nilpotent-by-polycyclic.

Theorems 11 and 11+ may be used to prove the following theorem.

(12) (Amberg, Franciosi, de Giovanni [21]). *Let the soluble group $G = AB$ of derived length n be the product of a nilpotent-by-polycyclic subgroup A and a polycyclic subgroup B. Then G is nilpotent-by-polycyclic and, if G is not abelian, the torsion-free rank of the Fitting factor group of G satisfies*

$$r_0(G/\text{Fit } G) \le (2n - 3)(r_0(B) + r_0(A/\text{Fit } A)).$$

Comments on the proof.

The proof of the rank inequalities uses Theorem 11+ as well as some facts about the automor-phism group of finitely generated abelian groups that depend ultimately on algebraic number theory, in particular on Dirichlet's Unit Theorem. (For the case when A and B are abelian, see Heineken [52]). The following proof of the fact that G is nilpotent-by-polycyclic is easier and depends mainly on Theorem 11:

Assume that G is not nilpotent-by-polycyclic. Let $G = AB$ be a counterexample with minimal derived length and let $K \ne 1$ be the last term of the derived series of G. Then there exists a nilpotent normal subgroup L/K of G/K with polycyclic factor group G/L. Let $F = $ Fit A be the Fitting subgroup of A and c the nilpotency class of F.

1. *The case $F \cap K = 1$.*

In this case

$$A \cap K \simeq (A \cap K)F/F \subseteq A/F \quad \text{and} \quad (A \cap BK)/(A \cap K) \simeq (A \cap BK)K/K \subseteq BK/K$$

are polycyclic, so that also $A \cap BK$ is polycyclic. By Theorem 11 the factorized group $BK = B(A \cap BK)$ is also polycyclic. In particular K is finitely generated. As a soluble group of automorphisms of a polycyclic group also $G/C_G(K)$ is polycyclic (see [68], Part 1, p.82). Then also $G/(L \cap C_G(K))$ is polycyclic. Since K is contained in the center of $L \cap C_G(k)$ and $(L \cap C_G(K))/K \subseteq L/K$ is nilpotent, also $L \cap C_G(K)$ is nilpotent. Therefore G is nilpotent-by-polycyclic.

2. *The general case.*

Note that $F \cap K$ is normal in AK and that $F/(F \cap K)$ is a nilpotent normal subgroup of $A/(F \cap K)$ with polycyclic factor group A/F. Consider the factorized group $AK = A(B \cap AK)$ and its factor

group
$$AK/(F \cap K) = (A/(F \cap K))((B \cap AK)(F \cap K)/(F \cap K)).$$

By the first case $AK/(F \cap K)$ and also its subgroup $FK/(F \cap K)$ are nilpotent-by-polycyclic. Since $F \cap K = Z_c(F) \cap K$ is contained in the c-th term $Z_c(FK)$ of the upper central series of FK, also FK is nilpotent-by-polycyclic. Since $AK/FK \simeq A/(A \cap FK)$ is polycyclic, also AK is nilpotent-by-polycyclic.

Put $H = AK \cap L$. Then $K \subseteq H \subseteq L$ and since L/K is nilpotent, H is a nilpotent-by-polycyclic subnormal subgroup of G. Let

$$H = H_0 \lhd H_1 \lhd H_2 \lhd \cdots \lhd H_t = G$$

be the standard series of H in G. Since A normalizes H, it is well-known that A also normalizes each H_i. Then $A \subseteq N_G(H_i)$ and so $N_G(H_i) = A(B \cap N_G(H_i))$. Consider the factorized group

$$N_G(H_i)/H_i = (AH_i/H_i)((B \cap N_G(H_i))H_i/H_i).$$

The group AH_i/H_i is an epimorphic image of

$$A/(A \cap H) = A/(A \cap L) \simeq AL/L \subseteq G/L,$$

and hence it is polycyclic. By Theorem 11 the group $N_G(H_i)/H_i$ is also polycyclic. Hence, if H_i is nilpotent-by-polycyclic for some i, also $N_G(H_i)$ and its subgroup H_{i+1} are nilpotent-by-polycyclic. Since $H_0 = H$ is nilpotent-by-polycyclic, also $G = H_t$ is nilpotent-by-polycyclic.

Remarks. (a) For non-soluble groups Theorem 12 becomes false in general. This can be seen from the alternating group of degree 5 which can be written as a product $G = AB$ where A is a cyclic group of order 5 and B is an alternating group of degree 4, but G is not soluble.

(b) Let p be a prime and $A = \operatorname{Aut} B$ be the automorphism group of a Prüfer group B of type p^∞. Then A is an uncountable abelian group. Consider the holomorph $G = AB$. Then $B = \operatorname{Fit} G$ and $G/\operatorname{Fit} G$ is not a Chernikov group, since it is uncountable. Therefore in Theorem 12 'polycyclic' cannot be replaced by 'Chernikov'. However, it is not difficult to show that a soluble product of a hypercentral-by-Chernikov group and a Chernikov group must be an extension of a locally nilpotent group by a Chernikov group.

Problem. Determine the structure of (soluble) groups $G = AB$ where A is a subgroup satisfying some generalized nilpotency condition and the subgroup B satisfies some finiteness condition.

VI. Minimax Groups.

A group G is a *minimax group* if it has a finite series G_i such that

$$1 = G_0 \lhd G_1 \lhd G_2 \lhd \cdots \lhd G_n = G$$

whose factors G_{i+1}/G_i satisfy the maximum or minimum condition on subgroups. It is easy to see that a soluble group G is a minimax group if it has such a series in which each factor G_{i+1}/G_i is cyclic or a Prüfer group of type p^∞ for some prime p. The number of infinite factors in any such series is an invariant of the group G which is called its *minimax rank* $m(G)$.

Problem. Is every product $G = AB$ of two (soluble) minimax subgroups A and B a minimax group?

If the group $G = AB$ is soluble, this question has a positive answer.

(13) (Wilson [88] and [90]). *If $G = AB$ is a soluble group where A and B are minimax subgroups of G, then G is a minimax group and the minimax rank of G satisfies*

$$m(G) = m(A) + m(B) - m(A \cap B).$$

The special case of this theorem, where A or B is nilpotent, was previously proved by Amberg and Robinson in [26] and by Zajtsev in [96]; the case when A and B are abelian was already treated in [93].

The proofs of these results reduce quickly to the situation of a triply factorized group

$$G = AB = AK = BK$$

where A and B are minimax subgroups and K is an abelian normal subgroup of G with $A \cap K = B \cap K = 1$.

Regard K as an A-module. The map δ from A to K assigning to each element a of A the unique element of $(a^{-1}B) \cap K$ is surjective and is easily checked to be a *derivation*, that is

$$(a_1 a_2)\delta = (a_1\delta)a_2 + a_2\delta \quad \text{for all } a_1, a_2 \text{ in } A.$$

Theorem 13 follows immediately from a fact proved in [90] that if δ is a surjective derivation from a soluble minimax group A to an A-module K, then K, regarded as a group, is minimax (This even holds when the derivation δ is only nearly surjective, that is the set $A\delta$ has a finite complement in K). To see this, after some reductions one has to show that there are no infinite simple modules K for a soluble minimax group A with the property that there is a (nearly) surjective derivation from A to K. Note that all soluble minimax groups which are not polycyclic have infinite simple modules, but apart from some ideas in a paper of Brookes [30] there is little known about these. (See however Theorem 19 below). In the proof also some cohomological results are used.

For the proof of the rank formula for the minimax rank in Theorem 13 number theoretical arguments such as Dirichlet's Unit Theorem are decisive.

Corresponding to Theorem 12 there is the following result. Recall that the *Gruenberg radical* $K(G)$ of the group G is the subgroup generated by all its abelian ascendant subgroups.

(14) (Amberg, Franciosi, de Giovanni [21]). *If the soluble group $G = AB$ of derived length n is the product of a hypercentral-by-minimax subgroup A and a minimax subgroup B, then G is an extension of its Gruenberg radical by a minimax group, and, if G is not abelian, the minimax rank of the Gruenberg factor group satisfies*

$$m(G/K(G)) \leq (2n - 3)(m(B) + m(A/H)),$$

where H is any hypercentral normal subgroup of A with minimax factor group A/H.

VII. Finite Rank.

There are several types of groups of finite rank which are considered.

(a) A group G has *finite torsion-free rank* if it has a series of finite length n in which each non-periodic factor is infinite cyclic. The number of infinite cyclic factors in any such series is an invariant of G called its *torsion-free rank* $r_0(G)$.

(b) A group G has *finite abelian section rank* if every elementary abelian section (p-factor) of G is finite for every prime p.

(c) A group G has *finite Prüfer rank* $r = r(G)$ if every finitely generated subgroup of G can be generated by r elements and r is the least such number.

For soluble groups the relations between the various classes are indicated in the following diagram. Simple examples show that even for abelian groups all the inclusions are proper.

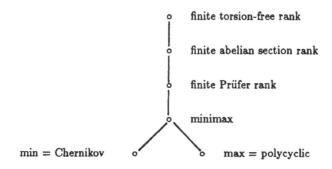

Problem. If the group $G = AB$ is the product of two (soluble) subgroups A and B of finite rank in some sense, does then G have finite rank in this same sense?

This problem was first considered in Amberg [4]. The example of Suchkov in Theorem 7 shows that an arbitrary product of two groups with finite torsion-free rank can have infinite torsion-free rank. On the other hand, Sysak has announced that every soluble product $G = AB$ of two subgroups A and B with finite torsion-free rank has likewise finite torsion-free rank. If A and B are abelian, this was known already from [93]. Robinson in [72] considered the case when A or B is nilpotent. Some more general special cases when A and B satisfies some weak nilpotency requirements were treated in Amberg [12], Chernikov [42] and Zajtsev [98]. Also it is shown in [25] that every factorized group $G = AB$ of finite torsion-free rank satisfies

$$r_0(G) \leq r_0(A) + r_0(B) - r_0(A \cap B).$$

It seems to be unknown at present whether in the general case we have an equality here. The difficulty is to prove that the rank equality holds for the torsion-free rank in the case that the group G has the form $G = AB = AT = BT$ where T is a periodic normal subgroup of G.

Every soluble group G with finite abelian section rank has a finite series

$$1 = G_0 \lhd G_1 \lhd G_2 \lhd \cdots \lhd G_t = G$$

whose factors G_{i+1}/G_i are infinite cyclic or abelian torsion groups whose primary components satisfy the minimum condition. The number of Prüfer p-groups in any such series is finite and an invariant of G which is called its p^∞-rank. The following theorem solves the above problem for soluble groups with finite abelian section rank.

(15) (Wilson [91]. *If the soluble group $G = AB$ is the product of two subgroups A and B of finite abelian section rank, then G has finite abelian section rank.*

The special case of this when A and B are abelian was treated by Zajtsev in [93], Robinson considered the case when A or B is nilpotent in [72]. Some further more general special cases of Theorem 15 can be found in Amberg [4] and [13] and Chernikov[44]. Using the arguments of Wilson [88] it is possible to prove the following additional information on the structure of the factorized group in Theorem 15 (see Amberg and Müller [25]).

(15+) *If $G = AB$ is a soluble group with finite abelian section rank, then for every prime p and for $p = 0$ the ranks $r_p(G)$ satisfy*

$$r_p(G) = r_p(A) + r_p(B) - r_p(A \cap B).$$

From Theorem 15 it follows easily that also every soluble product $G = AB$ of two subgroups A and B with finite Prüfer rank likewise has finite Prüfer rank (see [13] or [91]). Morover, the Prüfer rank of G is bounded by a function of the Prüfer ranks of A and B (see [13]). For finite G this was already shown by Zajtsev in [95] (see also Robinson [72] and Amberg and Robinson [26]).

Again the proofs of these results involve cohomological methods as well as some facts about finitely generated modules over locally almost-polycyclic groups: see Theorem 19 below. In the proof of Theorem 15 theorems about surjective derivations are decisive.

VII. Auxiliary Results.

In this section we mention some auxiliary results that are of importance in the proofs of the theorems mentioned earlier. The first two elementary lemmas have proved useful in the study of periodic factorized groups.

(16) (Wielandt [87]; see Kegel [61]). *Let the group $G = AB$ be factorized by two subgroups A and B one of which is periodic. If A^* is a normal subgroup of A and B^* is a normal subgroup of B, then the normalizer $N_G(\langle A^*, B^* \rangle)$ is factorized.*

(17) (Amberg [2] and [3]). *Let the group $G = AB$ be factorized by two subgroups A and B. If the subgroup A^* has finite index in A and the subgroup B^* has finite index in B, then also $\langle A^*, B^* \rangle$ has finite index in G.*

Also the next lemma on triply factorized groups is elementary and is useful whenever one of the two subgroups A or B has a non-trivial center.

(18) (Amberg [12]). *Let $G = AB = AK = BK$, where A and B are subgroups and K is a non-trivial abelian normal subgroup of the group G with $A \cap K = B \cap K = 1$. If $a \in Z(A)$, then the commutator subgroup $[K, a]$ is normal in G and properly contained in K.*

Since our knowledge about finitely generated modules over groups with finite rank is rather meager the following result is invaluable.

(19) (Zajtsev [93]). *Let Q be a locally almost-polycyclic group with finite torsion-free rank, $\mathbf{J}$ be a principal ideal domain and M a finitely generated $\mathbf{J}Q$-module. If $M = Mf$ for infinitely many primes f in $\mathbf{J}$, then M is a $\mathbf{J}$-torsion module.*
(This implies for instance that a simple $\mathbb{Z}Q$-module must be an elementary abelian p-group).

Finally we mention the following cohomological result.

(20) (Robinson [73]). *Let G be a locally nilpotent group, $\mathbf{R}$ a ring with identity and M an $\mathbf{R}G$-module such that $G/C_G(M)$ is a hypercentral group. Then the cohomology and the homology groups satisfy $H^n(G, M) = 0 = H_n(G, M)$ for all $n \geq 0$ provided that either* (i) *M is $\mathbf{R}G$-noetherian and $H_0(G, M) = 0$ or* (ii) *M is $\mathbf{R}G$-artinian and $H^0(G, M) = 0$.*

IX. Triply Factorized Groups.

If K is a normal subgroup of the factorized group $G = AB$, then the factorizer $X(K) = AK \cap BK$ has the triple factorization

$$X(K) = K(A \cap BK) = K(B \cap AK) = (A \cap BK)(B \cap AK).$$

Therefore it is of interest to study triply factorized groups of the form

$$G = AB = AK = BK \quad \text{with } K \text{ normal in } G.$$

If the group theoretical property $\mathfrak{X}$ is inherited by epimorphic images and extensions, then clearly the triply factorized group G is an $\mathfrak{X}$-group whenever the subgroups A, B and K are $\mathfrak{X}$-groups. This will usually not be the case if $\mathfrak{X}$ is some nilpotency requirement. On the other hand, it is for instance shown in Amberg [6] that a locally finite group $G = AB = AK = BK$, where A and B are hypercentral subgroups and K is locally nilpotent normal subgroup of G, is locally nilpotent.

In [81] Sysak gives an example of a triply factorized group $G = AB = AK = BK$ where A, B and K are torsion-free abelian and K is normal in G with Prüfer rank 1, but G is not nilpotent in any reasonable sense; in particular G is not locally nilpotent. However, if K is only a minimax group the following general theorem can be proved.

(21) (Amberg, Fanciosi, de Giovanni [16], [17] and [22]). *Let the group $G = AB = AK = BK$ be the product of two subgroups A and B and a normal minimax subgroup K. If A, B and K satisfy some nilpotency requirement $\mathfrak{N}$, then G satisfies the same nilpotency requirement $\mathfrak{N}$. Here $|GN$ can for instance be chosen to be one of the following group classes:*
    (i) *the classes of nilpotent, hypercentral or locally nilpotent groups,*
    (ii) *the classes of FC-nilpotent and FC-hypercentral groups,*
    (iii) *The classes of nilpotent-by-$\mathfrak{X}$, hypercentral-by-$\mathfrak{X}$ and (locally nilpotent)-by-$\mathfrak{X}$ groups, where $\mathfrak{X}$ is the class of finite, periodic, polycyclic, Chernikov or minimax groups.*

For the class $\mathfrak{N}$ of nilpotent-by-finite groups as well as for the class of finite-by-nilpotent groups the statement of Theorem 21 was first proved by Halbritter in [51]. – It can also be shown that if A and B are supersoluble (hypercyclic, locally supersoluble) and K is a hypercentral minimax group, then G is supersoluble (hypercyclic, locally supersoluble) (see [18]) – Similar statements as those in Theorem 21 -except in the case when A, B and K are nilpotent where it is in addition needed that the maximal periodic (normal) subgroup $T(K)$ of K is a Chernikov group – hold if the group G has finite abelian section rank (and K is not necessarily a minimax group).

For the proof of Theorem 21 and the related results one reduces first to the case where the normal subgroup K is abelian. (For this so-called *theorems of Hall's type* in the sense of [68], p.37, had to be developed). Then again cohomology theory is applied and in particular Theorem 20 is used.

In [62] Kegel has shown that a 'trifactoized' finite group $G = AB = AC = BC$ where A, B and C ar nilpotent subgroups, is likewise nilpotent (for another proof of this see Pennington [66]). Some extensions of this theorem to almost soluble minimax groups can be found in Amberg, Franciosi and de Giovanni [19] and Amberg and Halbritter [23].

As an application of the statement corresponding to Theorem 21 that a soluble group $G = AB = AK = BK$ with finite abelian section rank is hypercentral whenever A, B and K are hypercentral, we prove the following corollary. The *Hirsch-Plotkin radical* $R = R(G)$ of a group G is the product of all its locally nilpotent normal subgroups, $R(G)$ is itself a locally nilpotent characteristic subgroup of G by a famous theorem of Hirsch and Plotkin (see [70], p.343). It was noted in Amberg [2] and [3] that the Hirsch-Plotkin radical of a product $G = AB$ of two locally nilpotent subgroups, where G is polycyclic or a Chernikov group, is likewise factorized. (For finite groups see also Pennington [65]). The following more general fact can now be proved (see [16]).

(22) Corollary. *Let the soluble group $G = AB$ be the product of two hypercentral subgroups A and B with finite abelian section rank. Then G has finite abelian section rank (by Theorem 15) and the Hirsch-Plotkin radical R of G is factorized, i.e $R = (A \cap R)(B \cap R)$ and $A \cap B \subseteq R$.*

Proof. Consider the ascending Hirsch-Plotkin series

$$1 = R_0 \subseteq R_1 \subseteq \cdots \subseteq R_t = G$$

of G, where R_{i+1}/R_i is the Hirsch-Plotkin radical of G/R_i. For every $i \leq t$ let X_i be the factorizer of R_i in G. Then for every $i < t$ the subgroup X_{i+1}/R_i is the factorizer of R_{i+1}/R_i, so that X_{i+1}/R_i is hypercentral by the statement corresponding to Theorem 21 for a group with finite abelian section rank and three hypercentral factors. Since $R_i \subseteq X_i \subseteq X_{i+1}$, the subgroup X_i is ascendant in X_{i+1}. It follows that $X_1 = X(R_1)$ is a hypercentral ascendant subgroup of G, so that the Hirsch-Plotkin radical $R = X(R)$ of G is factorized.

The symmetric group of degree 4 has a factorization $G = AB$ where A is a symmetric group of degree 3 and B is a cyclic group of order 4, but the Hirsch-Plotkin radical $R(G) = \text{Fit } G$ is not factorized. This shows that in the corollary both subgroups A and B have to satisfy some nilpotency requirement.

More about factorized normal subgroups of a factorized group can for instance be found in Amberg [6], [7] and [10]; see also [22]. In particular for products of two abelian subgroups the following can be said.

(22) (Amberg [10] and [7]). *Let the group $G = AB$ be the product of two abelian subgroups A and B.*

(a) *Each term of the upper central series and each term of the upper FC-central series of G is factorized; in particular the centre, the hypercentre, the FC-centre and the FC-hypercentre of G are factorized.*

(b) *If G is not cyclic of prime order and $A \neq G$ or $B \neq G$, then there exists a factorized normal subgroup N of G with $1 \neq N \neq G$.*

(c) *If A and B are periodic, then the Hirsch-Plotkin radical of G is factorized.*

Problems. (a) Does the corollary still hold when $G = AB$ has only finite torsion-free rank?

(b) Find other 'factorized' normal subgroups N of a given factorized group $G = AB$ which satisfy

$N = (A \cap N)(B \cap N)$.

(c) Find other group theoretical properties $\mathfrak{X}$ which are inherited by triply factorized groups or even trifactorized groups $G = AB = AC = BC$ (possibly under additional conditions).

X. Special Normal Subgroups.

If the group $G = AB$ is the product of two subgroups A and B, it is of some interest to know whether one of the following two conditions holds:

(a) There exists a non-trivial normal subgroup of $G \neq 1$ which is contained in A or B.

(b) There exists a proper normal subgroup of G containing A or B.

In general, neither of the two conditions must be satisfied. For instance, the alternating group of degree 5, $G = A_5$, is factorized by a (soluble) subgroup which is isomorphic to the alternating group of degree 4 and a cyclic group of order 5, but G does not contain any normal subgroup other than 1 or G. Moreover, Holt and Howlett in [54] have constructed metabelian p-groups $G = AB$ where A and B are infinite elementary abelian p-subgroups of G, but G does not satisfy (a) nor (b).

Already in [57] Itô showed that (a) holds for every finite group $G \neq 1$ which is the product of two abelian subgroups A and B. Cohn proved in [49] that (a) also is valid for products of two cyclic groups. Sesekin in [76] extended Itô's result to arbitrary products $G = AB$ of two abelian subgroups A and B, one of which satisfied the minimum condition on subgroups. Around 1972 Amberg and independently Sesekin proved that here the minimum condition can be replaced by the maximum condition on subgroups (see [77] and [2]). Some further results about (a) are contained in Amberg [11]. The most far reaching theorem about this problem is the following.

(24) (Zajtsev [93]). *If the group $G = AB$ is the product of two abelian subgroups A and B, one of which has finite section rank, then there exists a normal subgroup $N \neq 1$ of $G \neq 1$ which is contained in A or B.*

In [57] Itô showed that condition (b) holds for finite products of two different abelian groups A and B. This was extended by Kegel in [60] to finite products of two different nilpotent groups. Amberg showed in [5] that (b) holds for every soluble product $G = AB$ of two different nilpotent subgroups A and B, where A or B satisfies the maximum or minimum condition for subgroups. This can be easily extended to the case when A or B is a minimax group (see Amberg and Robinson [26]). The most general theorem about condition (b) is the following.

(25) (Amberg, Franciosi, de Giovanni [15]). *Let the soluble group $G = AB$ be factorized by two different nilpotent subgroups A and B. If at least one of the two subgroups A and B has finite abelian section rank, then there exists a proper normal subgroup of G containing A or B.*

References

[1] B. Amberg, 'Abelian factorizations of infinite groups', *Math. Z.* 123(1971), 201-214.

[2] B. Amberg, 'Factorizations of infinite groups', Habilitationsschrift, Univerität Mainz, Mai 1973.

[3] B. Amberg, 'Artinian and noetherian factorized groups', *Rend. Sem. Mat. Univ. Padova* 55(1976), 105-122.

[4] B. Amberg, 'Factorizations of infinite soluble groups', *Rocky Mountain J. Math* 7(1977), 1-17.

[5] B. Amberg, 'Über auflösbare Produkte nilpotenter Gruppen', *Arch. Math.* (Basel) 30(1978), 361-363.

[6] B. Amberg, 'Lokal endlich-auflösbare Produkte von zwei hyperzentralen Gruppen', *Arch. Math.* (Basel) 35(1980), 228-238.

[7] B. Amberg, 'Über den Satz von Kegel und Wielandt', *Arch. Math.* (Basel) 40(1983), 289-296.

[8] B. Amberg, 'Soluble products of two locally finite groups with min-p for every prime p', *Rend. Sem. Mat. Padova* 69(1983), 7-17.

[9] B. Amberg, 'Factorized groups with max, min, and min-p', *Can. Math. Bull.* 27(1984), 171-178.

[10] B. Amberg, 'Products of two abelian subgroups" *Rocky Mountain J. Math.* 14(1984), 541-547.

[11] B. Amberg, 'On groups which are the product of abelian subgroups', *Glasgow Math. J.* 26(1985), 151-156.

[12] B. Amberg, 'Produkte von Gruppen mit endlichem torsionsfreiem Rang', *Arch. Math.* (Basel) 45(1985), 398-406.

[13] B. Amberg, 'Products of groups with finite rank', *Arch. Math.* (Basel) 49(1987), 369-375.

[14] B. Amberg, 'Infinite factorized groups: results and problems', Workshop on Algebra and Number Theory, Pusan, South Korea, 1987.

[15] B. Amberg, S. Franciosi and F de Giovanni, 'On normal subgroups of products of nilpotent groups', *J. Austral Math. Soc.* (Ser. A) 44(1988), 275-286.

[16] B. Amberg, S. Franciosi and F de Giovanni, 'Groups with a nilpotent triple factorization', *Bull. Austral Math. Soc.* 37(1988), 69-79.

[17] B. Amberg, S. Franciosi and F de Giovanni, 'Groups with an FC-nilpotent triple factorization', *Ricerche Mat.*, to appear.

[18] B. Amberg, F. Franciosi and F de Giovanni, 'Groups with a supersoluble triple factorization', *J. Algebra* 117(1988), 136-148.

[19] B. Amberg, F. Franciosi and F de Giovanni, 'On trifactorized soluble minimax groups', *Arch. Math.* (Basel) 51 (1988), 13-19.

[20] B. Amberg, S. Franciosi and F de Giovanni, 'Soluble groups which are the product of a nilpotent and polycyclic subgroup', *Proceedings of the Singapore Group Theory Conference*, (1987).

[21] B. Amberg, S. Franciosi and F de Giovanni, 'Nilpotent-by-noetherian factorized groups', *Canad. Math. Bull.*, to appear.

[22] B. Amberg, S. Franciosi and F de Giovanni, 'Triply factorized groups', to appear.

[23] B. Amberg and N. Halbritter, 'On groups with a triple factorization', to appear.

[24] B. Amberg and R. Moghaddam, 'Products of locally finite groups with min-p', *J. Austral Math. Soc.* (Ser. A) 41(1968), 352-360.

[25] B. Amberg und T. Müller, 'Rangfunktionen für Produkte von Untergruppen', to appear.

[26] B. Amberg and D.J.S. Robinson, ' Soluble groups which are products of nilpotent minimax groups', *Arch. Math.* (Basel) 42(1984), 385-390.

[27] B. Amberg and W.R. Scott, 'Products of abelian subgroups', *Proc. Amer. Math. Soc.* 26(1970), 541-547.

[28] M. Blaum, 'Factorizations of the simple groups PSL $(3,q)$ and $PSU(3,q^3)$', *Arch. Math.* 40(1983), 8-13.

[29] W. Brisley and I.D. MacDonald, 'Two classes of metabelian p-groups', *Math. Z.* 112(1969), 5-12.

[30] C.J.B. Brookes, 'Ideals in group rings of soluble groups of finite rank', *Math. Proc. Camb. Phil. Soc.* 97(1985), 27-49.

[31] N.S. Chernikov, 'Infinite groups that are products of nilpotent subgroups', *Dokl. Akad. Nauk SSSR* 252(1980), 57-60.

[32] N.S. Chernikov, 'Groups which are factorized by extremal subgroups', *Ukrain. Mat. Zh.* 32(1980), 707-711.

[33] N.S. Chernikov, 'Some factorization theorems for infinite groups', *Dokl. Akad. Nauk SSSR* 255(1980), 537-539.

[34] N.S. Chernikov, 'On factorized locally finite groups', *Sibir. Mat. Zh.* 21(1980), 186-195.

[35] N.S. Chernikov, 'Products of almost abelian groups', *Ukrain. Mat. Zh* 33(1981), 136-138.

[36] N.S. Chernikov, 'Products of groups of finite rank', *Algebra i Logika* 20(1981), 315-329.

[37] N.S. Chernikov, 'On particular factorization of locally finite groups', *A study of groups with given characteristic system of subgroups*, Akad. Nauk Ukrain. SSSR, Inst. Mat., Kiev (1981), 25-34.

[38] N.S. Chernikov, 'Factorized locally graduated groups', *Dokl. Akad. Nauk SSSR* 260(1981), 543-546.

[39] N.S. Chernikov, 'Locally graduated groups factorized by subgroups of finite rank', *Algebra i Logika* 21(1982), 108-120.

[40] N.S. Chernikov, 'Sylow subgroups of factorized periodic linear groups', *A subgroup characterization of groups*, Akad. Nauk Ukrain. SSSR, Inst. Mat., Kiev (1982), 35-52.

[41] N.S. Chernikov, 'On factorized locally finite groups satisfying the minimal condition on primary subgroups', *A subgroup characterization of groups*, Akad. Nauk Ukrain. SSSR, Inst. Mat., Kiev (1982), 52-73.

[42] N.S. Chernikov, 'Factorization theorem for locally graduated groups', *Ukrain. Mat. Zh.* 34(1982), 732-737.

[43] N.S. Chernikov, 'Infinite groups decomposed as a product of pairwise permutable subgroups', *Dokl. Akad. Nauk SSSR* Ser. A 11(1983), 24-27.

[44] N.S. Chernikov, 'Products of groups of finite free rank', *Groups and systems of their subgroups*, Akad. Nauk Ukrain. SSSR, Inst. Mat., Kiev (1983), 42-56.

[45] N.S. Chernikov, 'Factorizations of infinite groups in pairwise permutable subgroups', *Construction of groups and their subgroup characterization*, Akad. Nauk Ukrain. SSSR, Inst. Mat., Kiev (1984), 47-66.

[46] N.S. Chernikov, 'Factorizations of infinite groups under finiteness conditions', *Dokl. Akad. Nauk SSSR* Ser. A 5(1985), 26-29.

[47] N.S. Chernikov, 'Properties of the normal closure of the centre of an FC-subgroup B of a group $G = AB$ with abelian subgroup A', *Ukrain. Mat. Zh.* 36(1986), 364-368.

[48] N.S. Chernikov, 'Groups which are products of permutable subgroups', Akad. Nauk Ukrain. SSSR, Inst. Mat., Kiev (1987).

[49] P.M. Cohn, 'A remark on the general product of two infinite cyclic groups', *Arch. Math.* (Basel) 7(1956), 94-99.

[50] E. Fisman and Z. Arad, 'A proof of Szep's conjecture on non-simplicity of certain finite groups', *J. Alg.* 108(1987), 340-354.

[51] N. Halbritter, 'Groups with a nilpotent-by-finite triple factorization', *Arch. Math.* (Basel), to appear.

[52] H. Heineken, 'Produkte abelscher Gruppen und ihre Fittinggruppe', *Arch. Math.* (Basel), 48(1987),185-192.

[53] H. Heineken and J.C. Lennox, 'A note on products of abelian groups', *Arch. Math.* (Basel), 41(1983), 498-501.

[54] D.F. Holt and R.B. Howlett, 'On groups which are the product of two abelian groups', *J. London Math. Soc.* 29(1984), 453-461.

[55] R.B. Howlett, 'On the exponent of certain factorizable groups', *J. London Math. Soc.* 31(1985), 265-271.

[56] B. Huppert, *Endliche Gruppen I*, Springer, Berlin (1967).

[57] N. Itô, 'Über das Produkt von zwei abelschen Gruppen', *Math. Z.* 62(1955), 400-401

[58] N. Itô, 'On the factorizations of the linear fractional group LF(2,p)', *Acta. Math. Soc.* (Szeged) 15(1953), 79-84.

[59] L.S. Kazarin, 'On a problem of Szep', *Mat. USSR Izvestiya* 28(1987), 467-495.

[60] O.H. Kegel, 'Produkte nilpotenter Gruppen', *Arch. Math.* (Basel) 12(1961), 90-93.

[61] O.H. Kegel, 'On the solubility of some factorized linear groups', *Illinois J. Math.* 9(1965), 535-547.

[62] O.H. Kegel, 'Zur Struktur mehrfach faktorisierter endlicher Gruppen', *Math. Z.* 87(1965), 42-48.

[63] O.H. Kegel and B.A.F. Wehrfritz, *Locally finite groups*, North-Holland, Amsterdam (1973).

[64] J.C. Lennox and J.E. Roseblade, 'Soluble products of polycyclic groups', *Math. Z.* 170(1980), 153-154.

[65] E. Pennington, 'On products of finite nilpotent groups', *Math. Z.* 134(1973), 81-83.

[66] E. Pennington, 'Trifactorizable groups', *Bull. Austral. Math. Soc.* 8(1973), 461-469.

[67] L. Rédei, 'Zur Theorie der faktorisierbaren Gruppen. I.', *Acta Math. Hung.* 1(1950), 74-98.

[68] D.J.S. Robinson, *Finiteness conditions and generalized soluble groups, Part I and II*, Springer, Berlin (1972).

[69] D.J.S. Robinson, 'The vanishing of certain homology and cohomology groups', *J. Pure Appl. Algebra* 7(1976), 145-167.

[70] D.J.S. Robinson, *A course in the theory of groups*, Springer, Berlin (1980).

[71] D.J.S. Robinson, 'Infinite factorized groups', *Rend. Sem. Mat. Fis. Milano* 53(1983), 347-355.

[72] D.J.S. Robinson, 'Soluble products of nilpotent groups', *J. Algebra* 98(1986), 183-196.

[73] D.J.S. Robinson, 'Cohomology of locally nilpotent groups', *J. Pure Appl. Algebra* 48(1987), 281-300.

[74] J.E. Roseblade, 'Group rings of polycyclic groups', *J. Pure Appl. Algebra* 3(1973), 307-328.

[75] W.R. Scott, *Group Theory*, Prentice-Hall, England Cliffs, N.J. (1964).

[76] N.F. Sesekin, 'On the product of two finitely connected abelian groups', *Sibir Mat. Z.* 9(1968), 1427-1430.

[77] N.F. Sesekin, 'On the product of two finitely generated abelian groups', *Mat. Zametki* 13(1973), 266-268.

[78] S.E. Stonehewer, 'Subnormal subgroups of factorized groups', Lecture Notes in Math. 1281, *Proceedings Brixen/Bressanone* (1986), 158-175.

[79] N.M. Suchkov, 'Example of a mixed group factorized by two periodic subgroups', *Algebra i Logika* 23(1984), 573-577.

[80] N.M. Suchkov, 'Subgroups of a product of locally finite groups', *Algebra i Logika* 24(1985), 408-413.

[81] Y.P. Sysak, 'Products of infinite groups', Preprint 82.53, *Akad. Nauk Ukrain. Inst. Mat.*, Kiev (1982), 1-36.

[82] Y.P. Sysak, 'Products of locally cyclic, torsion-free groups', *Algebra i Logika* 25(1986), 672-686.

[83] H. Thiel, 'Die Faktorisationen der Gruppen PSL(2,K), PSL(3,K) und PSU(3,K) für einen lokal endlichen Körper', Diplomarbeit, Universität Mainz (1983).

[84] M.J. Tomkinson, 'Products of abelian subgroups', *Arch. Math.* (Basel) 47(1986), 107-122.

[85] H. Völklein, 'Algebraic groups as products of two subgroups, one of which is solvable', *Houston J. Math.* 12(1986), 145-154.

[86] H. Wielandt, 'Über das Produkt von paarweise vertauschbaren nilpotenten Gruppen', *Math. Z.* 65(1951), 1-7.

[87] H. Wielandt, 'Über Produkte nilpotenter Gruppen', *Illinois J. Math.* 2(1958), 611-618.

[88] J.S. Wilson, 'On products of soluble groups of finite rank', *Comment. Math. Helv.* 60(1985), 337-353.

[89] J.S. Wilson, 'Soluble groups which are products of minimax groups', *Arch. Math.* (Basel) 50(1988), 193-198.

[90] J.S. Wilson, 'Soluble products of minimax groups and nearly surjective derivations', *J. Pure Appl. Algebra*, to appear.

[91] J.S. Wilson, 'Soluble groups which are products of groups of finite rank', to appear.

[92] J.S. Wilson, 'A note on products of abelian-by-finite groups', *Arch. Math.* (Basel), to appear.

[93] D.I. Zajtsev, 'Products of abelian groups', *Algebra i Logika* 19(1980), 150-173.

[94] D.I. Zajtsev, 'Nilpotent approximations of metabelian groups', *Algebra i Logika* 20(1981), 638-653.

[95] D.I. Zajtsev, 'Factorizations of polycyclic groups', *Mat. Zametki* 29(1981), 481-490.

[96] D.I. Zajtsev, 'Products of minimax groups', *Groups and systems of their subgroups*, Akad. Nauk Ukrain. SSSR, Inst. Mat., Kiev (1983), 15-31.

[97] D.I. Zajtsev, 'Itô's theorem and products of groups', *Mat. Zametki* 33(1983), 807-818.

[98] D.I. Zajtsev, 'Soluble factorized groups', *Construction of groups and their subgroup characterization*, Akad. Nauk Ukrain. SSSR, Inst. Mat., Kiev (1984), 15-33.

[99] D.I. Zajtsev, 'Itô's theorem and products of groups', *Ukrain. Mat. Z.* 38(1986), 427-431.

Fachbereich Mathematik
Universität Mainz
D-6500 Mainz,
Federal Republic of Germany.

INFINITE GENERATION OF AUTOMORPHISM GROUPS

Seymour Bachmuth and H.Y. Mochizuki

Let $G(n)$ be the free group of finite rank n in a variety V. Since 1982, it has been known that Aut $G(n)$, the automorphism group of $G(n)$, may not be finitely generated for certain n. But is it always true that Aut $G(n)$ is finitely generated for all but a few number of dimensions n? Might this be the case if we restrict the variety V to be locally solvable? Until now, no examples to the contrary were known, and if we require $G(n)$ to be torsion free for almost all n, the questions remain unanswered. Below we shall exhibit varieties for which Aut $G(n)$ is infinitely generated for every $n \geq 2$. In order to place these results in context and discuss certain unsolved problems, we first discuss the state of knowledge concerning the automorphism groups.

If V is the variety of all groups so that $G(n)$ is the (absolutely) free group of rank n, then Aut $G(n)$ is finitely generated for all n [11]. If V is the variety of abelian groups so that $G(n)$ is the free abelian group of rank n, then Aut $G(n) \cong \mathrm{GL}_n(\mathbf{Z})$, the $n \times n$ invertible matrices over $\mathbf{Z}$, which is of course a finitely generated group. More generally, If V is any nilpotent variety, then it is well known and easy to show that Aut $G(n)$ is a finitely generated group for all n. For other varieties, prior to 1982, there was only the isolated result that the automorphism group of the free metabelian group of rank 2 is finitely generated [3]. Although our knowledge remains meager there has been some notable progress from 1982 to the present. The metabelian [5], Jacobi [7], and center–by–metabelian [12] varieties all have the property that Aut $G(3)$ is not finitely generated, but Aut $G(n)$ is finitely generated for all $n > 3$. Elena Stöhr has also shown in [12] that Aut $G(2)$ is not finitely generated for the center–by–metabelian variety. (It may also be worth pointing out the similar phenomena exhibited by another family of automorphism groups – the Torelli groups $\Im(n)$ of genus n, where $\Im(2)$ is not finitely generated ([9] and also [10]), but $\Im(n)$ is finitely generated for all $n > 2$ [8].)

The above examples, together with the stable range theorems of algebraic K–theory (and particularly the results of Suslin [13] on GL_n of polynomial and Laurent polynomial rings) lead us to ask the question below. Notice that in each of the metabelian, Jacobi, and center–by–metabelian varieties, $G(3)$ is torsion free yet Aut $G(3)$ is not finitely generated. The force of the hypothesis on V in the following question is to prohibit too much torsion in the $G(n)$.

Question: Let V be a solvable variety which does not satisfy an exponent law. Let $G(n)$ be the free group of rank n in V. Does there exist an integer $N(V)$ depending on V such that Aut $G(n)$ is finitely generated for all $n \geq N(V)$?

In the theorem below we shall produce examples of varieties in which the free groups $G(n)$

have the property that Aut $G(n)$ is not finitely generated for all $n \geq 2$. These varieties are not counterexamples for the question since we are permitting too much torsion in the $G(n)$.

Theorem: *Let k be an integer such that the ring $\mathbf{Z}/k\mathbf{Z}$ is not semisimple, and let $G(n)$ be the free group of rank n in the variety V of metabelian (solvable of length 2) groups whose commutator subgroups have exponent dividing k, i.e., V is defined by the two laws*

$$[[x,y],[u,v]] = 1 = [a,b]^k.$$

Then Aut $G(n)$ is not finitely generated for every $n \geq 2$.

<u>Sketch of Proof</u>: For simplicity, we shall assume that $k = p^2$, where p is a positive prime integer. Let $\mathbf{Z}_{p^2} = \mathbf{Z}/p^2\mathbf{Z}$ and $\mathbf{Z}_{p^2}A(n) = \mathbf{Z}_{p^2}[s_1, s_1^{-1}, \cdots, s_n, s_n^{-1}]$, the ring of Laurent polynomials in the n commuting indeterminates $s_1, \cdots, s_n$ over $\mathbf{Z}_{p^2}$. Since the commutator subgroup is a characteristic subgroup of $G(n)$, there is a natural map $\psi : \text{Aut } G(n) \longrightarrow \text{Aut}(G(n)_{ab}) \cong \text{GL}_n(\mathbf{Z})$, whose kernel we denote by $\text{IA}(G(n))$, i.e. we have the exact sequence

$$1 \longrightarrow \text{IA}(G(n)) \longrightarrow \text{Aut } G(n) \overset{\psi}{\longrightarrow} \text{Aut}(G(n)_{ab}) \longrightarrow 1.$$

Thus, Aut $G(n)$ acts on $\text{IA}(G(n))$, and it is sufficient to show that $\text{IA}(G(n))$ is not finitely generated as an Aut $G(n)$–group. Here, we are using the result: if N is a normal subgroup of a finitely generated group G such that G/N is finitely presented, then N is finitely generated as a G–group.

$\quad$ $\text{IA}(G(n))$ has a faithful representation ρ into $\text{GL}_n(\mathbf{Z}_{p^2}A(n))$, and Theorem 3.1 of [4] gives necessary and sufficient conditions for a matrix in $\text{GL}_n(\mathbf{Z}_{p^2}A(n))$ to be in the image of ρ. ρ extends to a faithful representation

$$\hat{\rho} : \text{Aut } G(n) \longrightarrow \text{GL}_n(\mathbf{Z}_{p^2}A(n)) \times \text{Aut}(G(n)_{ab}).$$

$\hat{\rho}$ is defined by

$$\hat{\rho} : \alpha \longmapsto ((a_{ij}), \psi(\alpha))$$

where $(a_{ij}) \in \text{GL}_n(\mathbf{Z}_{p^2}A(n))$ and $\sum_{i=1}^{n}(s_i - 1)a_{ij} = \psi(s_j) - 1$, $1 \leq j \leq n$. $\text{Aut}(G(n)_{ab})$ acts as a group of ring automorphisms on $\mathbf{Z}_{p^2}A(n)$ by extending its action on $A(n)$ linearly to $\mathbf{Z}_{p^2}A(n)$. Then $\text{Aut}(G(n)_{ab})$ acts on a matrix in $\text{GL}_n(\mathbf{Z}_{p^2}A(n))$ by acting on its entries. Note that $\text{Aut}(G(n)_{ab})$ also acts on $U(\mathbf{Z}_{p^2}A(n))$, the group of multiplicative units of $\mathbf{Z}_{p^2}A(n)$ and that each of the preceding actions of $\text{Aut}(G(n)_{ab})$ lifts to an action of Aut $G(n)$.

$\quad$ By composing ρ with the determinant map, $det : \text{GL}_n(\mathbf{Z}_{p^2}A(n)) \longrightarrow U(\mathbf{Z}_{p^2}A(n))$, we obtain the homomorphism

$$det \circ \rho : \text{IA}(G(n)) \longrightarrow U(\mathbf{Z}_{p^2}A(n)).$$

which turns out to be an Aut $G(n)$–map. By the extension of the ideas and arguments of Theorem C and its proof in [2], we can show that the image, $\text{Im}(det \circ \rho)$, of $det \circ \rho$ consists of all elements of the form $\prod_{i=1}^{n} s_i^{e_i}(1 + pa)$ where $e_i \in \mathbf{Z}$ and $a \in \mathbf{Z}_{p^2}A(n)$. Thus

$$\text{Im}(det \circ \rho) \cong A(n) \times [1 + p\mathbf{Z}_{p^2}A(n)],$$

which is an $\text{Aut}(G(n)_{ab})$–group. $[1 + p \ \mathbf{Z}_{p^2}A(n)]$ is isomorphic as an $\text{Aut}(G(n)_{ab})$–group to the additive group $(\mathbf{Z}_{p^2}A(n), +)$.

It is easy to verify that $(\mathbf{Z}_{p^2}A(n),+)$ is not finitely generated as an $\text{Aut}(G(n)_{ab})$-group, whence $IA(G(n))$ is not finitely generated as an $\text{Aut}\,G(n)$-group. Thus $\text{Aut}\,G(n)$ is not a finitely generated group.

Remark:

One can formulate the question in a "non-commutative stability" setting by asking if there is an integer $N(U)$ such that $\text{Aut}\,G(n)$ becomes "tame" for all $n \geq N(U)$, meaning that for $n \geq N(U)$, all elements of $\text{Aut}\,G(n)$ lift to automorphisms of the absolutely free group of rank n. These are much more difficult questions. For example, although finitely generated, it is not known if $\text{Aut}\,G(n)$, $n \geq 4$, is tame for the Jacobi and the center-by-metabelian varieties. On the other hand, for nilpotent varieties of class lager than three, $\text{Aut}\,G(n)$ is finitely generated for all n, but $\text{Aut}\,G(n)$ is not tame in this sense [1]. For nilpotent varieties this is clearly the wrong definition of tameness.

References

[1] S. Andreadakis, 'Automorphisms of free group and free nilpotent groups', *Proc. London Math. Soc.*, 15 (1965), 239–268.

[2] S. Bachmuth, G. Baumslag, J. Dyer, and H.Y. Mochizuki, 'Automorphism groups of 2-generator metabelian groups', to appear in *J. London Math. Soc.*

[3] S. Bachmuth, 'Automorphisms of free metabelian groups', *Trans. Amer. Math. Soc.*, 118(1965), 93–104.

[4] S. Bachmuth and H.Y. Mochizuki, 'Automorphisms of a class of metabelian groups II', *Trans. Amer. Math. Soc.*, (1967), 294–301.

[5] ―――――――――, 'The non-finite generation of Aut(G), G free metabelian of rank 3', *Trans. Amer. Math. Soc.*, 270(1982), 693–700.

[6] ―――――――――, 'Aut(F) ⟶ Aut(F/F″) is surjective for free group F of rank ≥ 4', *Trans. Amer. Math. Soc.*, 292(1985), 81–101.

[7] ―――――――――, 'The tame range of automorphism groups and GL_n', preprint.

[8] D. Johnson, 'The structure of Torelli group I: A finite set of generators for $\mathfrak{I}$', *Annals of Math.*, 118(1983),423–442.

[9] D. McCullough and A. Miller, 'The genus 2 Torelli group is not finitely generated', *Topology and Its Applications*, 22(1986), 43–49.

[10] G. Mess, personal communication.

[11] J. Nielson, 'Die Isomorphismengruppe der freien Gruppen', *Math. Ann.*, 91(1924), 169–209.

[12] E. Stöhr, *On automorphisms of free center-by-metabelian groups*, Akademic Der Wissenschaft Der DDR, Inst. Für Mathematik, Berlin 1985.

[13] A.A. Suslin, 'On the structure of the special linear group over polynomial rings', *Isv. Akad. Nauk.*, 11(1977), 221–238.

Department of Mathematics
University of California
Santa Barbara, California 93106
U. S. A.

NON-ORIENTABLE AND ORIENTABLE REGULAR MAPS

Peter Bergau and D. Garbe

Abstract. There do exist quite a lot of papers investigating regular maps on connected compact 2-manifolds, in the orientable case as well as in the non-orientable case (see the References). We want to bring about some new aspects, which also lead to interesting applications.

It is well known that the closed orientable surface of characteristic 2χ is a regular two-fold cover of the closed non-orientable surface of characteristic χ. Thus, every non-orientable regular (i.e. flag-transitive) map of genus h induces an orientable regular map of genus $g = h - 1$. But not every orientable map is induced, and different non-orientable regular maps can induce the same orientable map. We study this relation more closely and give several criteria by which one can easily get a classification of the non-orientable regular maps of genus h, if the list of the orientable ones of genus g is known. We construct all non-orientable regular maps having less than 6 faces.

Our and earlier results indicate that the orientable case seems to admit in a certain sense much more regular maps than the non-orientable case. This becomes plausible, if we take into consideration the crystallographic aspect. The group of the map (or a factor group) acts as a Euclidean crystallographic point group on a $2g$- or $(h-1)$-dimensional Z-lattice, if the genus is g or h, respectively. So the crystallographic restrictions are stronger in the non-orientable case.

Apart from the standard abelianizing procedure we can construct in the non-orientable case a normal subgroup N^{**} of the group of the universal tessellation, yielding the Z-lattice and giving a new method for proving certain groups to be infinite. We illustrate this by showing that the infiniteness of the group $(2,3,7;9)$ – which was proved by C.C. Sims [20] and J. Leech [16], [17] – is an immediate consequence of our results, as well as the new informations that $(2,3,7;13)$ and $(2,3,7;15)$ are infinite. (In fact: Using earlier findings of the second author, one can prove further new results on the groups $G^{p,q,r}$ and $(l,m,n;q)$). A more general and systematic exposition of our method will be given in a forthcoming publication [1].

1. Definitions and basic facts.

A partition of a connected compact 2-manifold M into finitely many, say N_2, simply-connected regions, called *faces*, by embedding a graph is called a *map* X on M. The edges (vertices) of the graph are called *edges* (*vertices*) of X. Selecting exactly one point from the interior of each face and of each edge (so-called *midpoint*) and connecting the midpoint of each face by arcs with the vertices of the face and with the midpoints of the edges of that face, one gets a triangulation of

X. We can take the set $F(X)$ of the "triangles" as a system of representatives of the flags of X. Incidence-preserving permutations of the set $F(X)$ are called *automorphisms* of X.

There are several concepts of regularity (see for instance [21], [5], [6], [23]). We adopt the classical (and rather strong) one: X is called *regular*, if its group G_X of automorphisms acts transitively on $F(X)$. A weaker concept is, for instance, *Th-regularity* (Threlfall [21]): Every face of X is a p-gon and at each vertex there meet exactly q faces. In this case the map can be denoted by the Schläfli-symbol $\{p, q\}$. We call a map *B-regular* (Brahana [2]), if there exists an automorphism R which cyclically permutes successive edges of a face and if there exists an automorphism S which cyclically permutes successive edges of a vertex which is incident with that face.

Every map $\{p, q\}$ induces a universal map $\{p, q\}^*$ on the universal covering space of M. $\{p, q\}^*$ can be metrically realized as a tessellation of the hyperbolic plane, of the Euclidean plane or of the sphere by congruent p-gons (see the figure in Section 3.). $G_{\{p,q\}^*}$ can be presented by

$$[p, q] := \langle \sigma_1, \sigma_2, \sigma_3 \; ; \; \sigma_1^2 = \sigma_2^2 = \sigma_3^2 = (\sigma_1\sigma_2)^p = (\sigma_2\sigma_3)^q = (\sigma_1\sigma_3)^2 = 1 \rangle.$$

So $G_{\{p,q\}^*}$ is a Coxeter-group of type $\circ \overset{p}{\text{———}} \circ \overset{q}{\text{———}} \circ$, σ_1, σ_2, σ_3 are reflections in the sides of a flag. Taking $R := \sigma_1\sigma_2$ and $S := \sigma_2\sigma_3$, one gets

$$[p, q]^+ := \langle R, \; S \; ; \; R^p = S^q = (RS)^2 = 1 >$$

as a subgroup of index 2 in $[p, q]$.

A compilation of old results due to Fricke, Klein, Dyck, Threlfall, Wilkie [9], [7], [21], [22] yields

Theorem 1 :
(i) (a) *If $\{p, q\}$ is a Th-regular map on a connected compact 2-manifold M, then the covering group N of $\{p, q\}^*$ over $\{p, q\}$ is a torsion-free subgroup N of finite index in $[p, q]$.*
(i) (b) *Let N be a torsion-free subgroup of finite index in $[p, q]$. Then the orbit space $\{p, q\}^*/N$ yields a Th-regular map $\{p, q\}$ on a connected compact 2-manifold M.*

Applying first (a) and then (b), the original $\{p, q\}$ is recovered by $\{p, q\}^*/N$. In both cases (a) and (b) we have:

M is orientable, iff $N \subset [p, q]$. The Euler characteristic of M is

$$\chi = -N_1 \frac{pq - 2p - 2q}{pq} \tag{1.1}$$

where N_1 denotes the number of edges. The automorphism group $G_{\{p,q\}}$ is isomorphic to $^{N*}[p, q]^N/_N$. Two subgroups N_1, N_2 yield the same map, iff they are conjugate in [p,q].

If M is orientable of genus g, then N has a presentation

$$\langle a_1, \; b_i \; (i = 1, 2, \cdots, g) \; ; \; \prod_{i=1}^{g} [a_i, b_i] = 1 \rangle.$$

If M is non-orientable of genus h, then N has a presentation

$$\langle a_i, \ (i = 1, 2, \cdots, h) \ ; \ \prod_{i=1}^{h} a_i^2 = 1 \rangle.$$

Let N be a torsion-free subgroup of finite index in $[p, q]$. Then

(ii) $\{p, q\} = \{p, q\}^* / N$ is B-regular, iff $N^+ := N \cap [p, q]^+$ is normal in $[p, q]^+$.

(iii) $\{p, q\} = \{p, q\}^* / N$ is regular, iff N is normal in $[p, q]$.

Corollary: *Every regular map is B-regular. Every B-regular map is Th-regular. Every non-orientable B-regular map is regular.*

Remark 1: In the literature B-regular maps are often called regular [2], [5], [10]; Wilson calls them rotary [24]. B-regular maps which are not regular are often called irreflexible [5]. The surface of smallest genus admitting B-regular maps which are not regular is the orientable closed surface of genus 7 [10]. There are exactly 3 such maps of genus 7 (leaving aside duals) [12].

In the following we restrict our considerations to regular maps. The following table surveys the classification of regular maps of small genus, as it has been carried out so far:

(χ Euler-characteristic; $\chi = 2 - 2g$ or $2 - h$; g (resp. h) the genus of the orientable (resp. non-orientable) closed surface).

	orientable			non-orientable	
χ	number of reg. maps	Reference	χ	number of reg. maps	Reference
2	5 platonic polyhedra		1	4	
	∞ dihedra, hosohedra			∞	
0	∞	[3]	0	0	[8],[5]
-2	10	[5]	-1	0	
-4	20	[19]	-2	4	[13]
-6	20	[10]	-3	6	[14]
-8	26		-4	6	[15]
-10	23		-5	4	[18]
-12	21	[12]	-6	2	

2. Induction of orientable regular maps by non-orientable ones.

Looking at the intersection $N^+ = N \cap [p, q]^+$, we see that every regular map $\{p, q\}_{n0,\chi}$ on a non-orientable surface of Euler characteristic χ *induces canonically* a regular map $\{p, q\}_{0,2\chi}$ on the orientable surface of characteristic 2χ. This is a well-known phenomenon, reflecting the fact that the orientable closed surface of characteristic 2χ is a smooth double cover of the non-orientable surface of characteristic χ [10; p.39]. We want to study this relation more closely.

Lemma 1: *Let $\{p,q\}_{o,\chi}$ be an orientable regular map of characteristic $\chi \leq 0$ defined by a torsion-free normal subgroup N in $[p,q]$. Then $\{p,q\}_{0,\chi}$ is canonically induced by a regular map $\{p,q\}_{n0,\chi/2}$, iff*

$$G_{\{p,q\}_0} \simeq G^+_{\{p,q\}_0} \times C_2, \tag{2.1}$$

where $G^+_{\{p,q\}_0} := [p,q]^+/N^+$ and C_2 cyclic of order 2.

    Proof. Let the group of $\{p,q\}_0$ be isomorphic to the direct product (2.1). Let N^+ be the normal subgroup which defines the map $\{p,q\}_0$ according to Theorem 1 (iii). By presupposition we have

$$[p,q]/N^+ \simeq [p,q]^+/N^+ \times C_2.$$

The cyclic group C_2 establishes a normal subgroup N in $[p,q]$ such that $N \not\subset [p,q]^+$ and $[N : N^+] = 2$. In $[p,q]$ only the conjugates of σ_1, σ_2, σ_3, RS and of the powers of R and S do have finite order. If N were not torsion-free, then at least one of the σ_i would be contained in N^+. Therefore $[p,q]/N$ would be dihedral or cyclic of order 2 or trivial. This is impossible, as $\chi \leq 0$. By Theorem 1 the normal subgroup N yields the inducing $\{p,q\}_{n0}$. Conversely: If $\{p,q\}_0$ is canonically induced by $\{p,q\}_{n0}$, it is evident that $G_{\{p,q\}_0}$ is isomorphic to the asserted direct product.

Corollary: *Let $\chi \leq 0$. If $\{p,q\}_{0,\chi}$ is canonically induced by a non-orientable regular map, then there exists an epimorphism*

$$C_{(p,2)} \times C_{(q,2)} \longrightarrow (G^+_{\{p,q\}_0})^{ab},$$

where $(G^+)^{ab}$ denotes the commutator factor group of G^+ and $(p,2)$ denotes the gcd of p and 2.

Remark 2: This corollary is very useful for the classification of non-orientable regular maps via that of orientable regular maps. From our list of the orientable maps of genus 7 [12] we see immediately that all, except $\{7,3\}$ (and its dual), violate the condition of the corollary. It is then easy to show that there exists exactly one $\{7,3\}_{n0,-6}$. It is defined by $N := \text{Ncl}_{[p,q]}\{(\sigma_1\sigma_2\sigma_3)^9\}$ [5; p.139]. ($\text{Ncl}_G\{\sigma\}$ denotes the normal closure of σ in G.) In the same way one can easily derive the results of the right-hand side of our table from those of the left-hand side. Therefore it can often be more economical to do the orientable case first.

Remark 3: Whereas every non-orientable regular map uniquely determines the induced orientable map, the converse is not true. For example, the two maps $\{6,4\}_{n0,-2}$ induce the unique $\{6,4\}_{0,-4}$. Nevertheless, as the table and our Theorem 2 suggest, there seem to exist in a certain sense much more orientable regular maps than non-orientable ones. The crystallographic view of Section 4 will give an elucidation of this phenomenon.

    According to Theorem 1 the automorphism group $G_{\{p,q\}}$ and the "rotation group" $G^+_{\{p,q\}}$ of a regular map are factor groups of $[p,q]$ and of $[p,q]^+$ respectively. Unless stated otherwise, we will denote throughout this paper in the groups G and G^+ the elements which represent the cosets σ_i, R or S likewise by σ_i, R and S, respectively.

Lemma 2: *Let $\{p,q\}_{0,2\chi}$ be an orientable regular map of characteristic $2\chi \leq 0$. Then $\{p,q\}_{0,2\chi}$ is canonically induced by a regular map $\{p,q\}_{n0,\chi}$, iff $R \longrightarrow R^{-1}$, $S \longrightarrow S^{-1}$ is an inner automorphism* of $G^+_{\{p,q\}_{0,2\chi}}$ induced by an involution of G^+.

Proof. Let $\{p,q\}_{o,2\chi}$ be canonically induced. By Lemma 1 its group is $[p,q]/N \simeq G^+ \times C_2$. The direct factor of order 2 is generated by $g\sigma_2$ (for a suitable $g \in G^+$). We have $g \neq 1$. $g\sigma_2 \rightleftharpoons g$ implies that g is an involution, and $g\sigma_2 \rightleftharpoons R, S$ implies that R, S are strongly real elements, in fact

$$R^g = R^{-1}, \quad S^g = S^{-1}, \quad g^2 = 1. \tag{2.2}$$

On the other hand: If $g \in G^+$ has the properties (2.2), then by analogous calculations we get $g\sigma_2 \rightleftharpoons R, S$, and we can apply Lemma 1 to prove the assertion.

Lemma 3: *Let $\{p,q\}_{0,2\chi}$ be an orientable regular map with rotation group G^+ such that $R \longrightarrow R^{-1}$, $S \longrightarrow S^{-1}$ is an inner automorphism and the order of the center $Z(G^+)$ is odd. Then $\{p,q\}_{0,2\chi}$ is induced canonically by exactly one non-orientable regular map.*

Proof. If $g \in G^+$ determines the inner automorphism $R \longrightarrow R^{-1}, S \longrightarrow S^{-1}$, then $g\sigma_2 \in C_G(G^+)$ and $g \rightleftharpoons \sigma_2$. $|g\sigma_2|$ is an even number. As $(g\sigma_2)^2 \in Z(G^+)$, we have $|g\sigma_2| \equiv 2 \mod 4$. So $(g\sigma_2)^{|g\sigma_2|/2} \in C_G(G^+) \setminus G^+$ and $G \simeq G^+ \times \langle(g\sigma_2)^{|g\sigma_2|/2}\rangle$. By Lemma 1 $\{p,q\}_{0,2\chi}$ is canonically induced. Assume that it is induced by two different non-orientable regular maps which are defined by normal subgroups N_1, N_2 of $[p,q]$. The product n_1n_2 of the corresponding central involutions of G belongs to G^+, and this contradicts the presupposition that $|Z(G^+)|$ is odd.

The next theorem illustrates the second part of Remark 3.

Theorem 2: *Non-orientable regular maps with abelian automorphism group do not exist. The B-regular maps $\{p,q\}_{0,2-2g}$ with abelian rotation group are exactly the following ones: $\{4g+2, 2g+1\}_2$ (and its dual), $\{4g, 4g\}_{1,0}$ and $\{2g+2, 2g+2\}_2$, where $\{p,q\}_r$ is defined by*

$$R^p = S^q = (RS)^2 = (RSR^{-1}S^{-1})^{r/2} = 1 \tag{2.3}$$

and $\{4g, 4g\}_{1,0}$ is defined by

$$R^{4g} = S^{4g} = (RS)^2 = R^{2g-1}S^{-1} = 1$$

(we use the denotion of Coxeter and Moser [5; p.111, 105]). These maps are even regular.

Proof. By application of Lemma 2 we get the first part of the assertion. If the B-regular $\{p,q\}_0$ has an abelian rotation group G^+, then there exists an epimorphism $G^+ \longrightarrow ([p,q]^+)^{ab}$. Thus: If p is even, then by $|G^+| = N_2p$ we have $N_2 \leq 2$. If p is odd, then $G^+ \simeq C_{(2p,q)}$, hence $\simeq C_{2p}$ or $\simeq C_p$. In any case $N_2 \leq 2$. An inspection of the list of the B-regular maps with one or two faces [10; Satz 6.2, 6.3] finishes the proof.

3. Construction of non-orientable regular maps.

In order to get all non-orientable regular maps $^{N_2}\{p,q\}_{no}$ with exactly N_2 faces, the following method (already used in the orientable case [10]) works fairly well, if N_2 is small. We know

$|G_{N_2(p,q)_{n0}}| = 2pN_2$. Let π be the canonical epimorphism $[p,q] \longrightarrow [p,q]/N \simeq G$. Necessarily $\langle \pi\sigma_1, \pi\sigma_2 \rangle \simeq D_p$, where D_p is the dihedral group of order $2p$. The enumeration of the cosets of $\langle \pi\sigma_1, \pi\sigma_2 \rangle$ in G yields a transitive permutation representation of G on N_2 letters. So, determine all transitive permutation representations ρ of $[p,q]$ on N_2 letters such that

$$|\rho([p,q])| = N_2|\rho(\langle \sigma_1, \sigma_2 \rangle)| \tag{3.1}$$

holds and $\rho(\langle \sigma_1, \sigma_2 \rangle)$ keeps the letter 1 fixed. So, determine for each representation ρ a presentation of $H := \rho^{-1}(\text{Stab}(1))$ using the Reidemeister-Schreier method. Look for all epimorphisms $\varphi : H \longrightarrow D_p$ for which $N := \text{Ker}\varphi$ is a torsion-free normal subgroup of $[p,q]$ and

$$N \not\subset [p,q]^+ \tag{3.2}$$

The set of all such N then yields all non-orientable regular maps $^{N_2}\{p,q\}_{n0}$.

S. Wilson [24] has developed a similar algorithm by translating the Reidemeister-Schreier-part into geometric language. He furthermore constructed a lot of regular maps. A.S. Grek [13] showed the non-existence of non-orientable regular maps for $N_2 = 1$ and $N_2 = 2$. We easily get Grek's results by our method. For instance, in the case $N_2 = 1$ necessarily $G \simeq [p,q]/\text{Ncl}\langle \sigma_3 W(\sigma_1, \sigma_2) \rangle$, $W(\sigma_1, \sigma_2)$ a word of even length. Hence $W(\sigma_1, \sigma_2) = (\sigma_1\sigma_2)^{p/2}$, and this is impossible. We now state

Theorem 3: *For $N_2 = 3, 5, 7$ the non-orientable regular maps with exactly N_2 faces are the semi-hosohedra $^{N_2}\{2, 2N_2\}_{n0,1}$ and the maps $^{N_2}\{4n+2, 2N_2\}_{n0,1-2n(N_2-1)}$ defined by*

$$N = \text{Ncl}_{[p,q]}\langle (\sigma_1\sigma_2)^{2n+1}(\sigma_3\sigma_2)^{N_2-1}\sigma_3 \rangle. \tag{3.3}$$

Proof. An inspection of the possibilities shows that for $N_2 = 3, 5, 7$ there exists only one transitive representation ρ fulfilling (3.1), namely

$$\begin{aligned} \rho\sigma_1 &= (1), \\ \rho\sigma_2 &= (23)(45)\cdots(N_2-1\ N_2), \\ \rho\sigma_3 &= (12)(34)\cdots(N_2-2\ N_2-1). \end{aligned} \tag{3.4}$$

The assertion is therefore proved by the following proposition.

Proposition: *If N_2 is an odd positive integer and if ρ is given by (3.4), then the algorithm described above yields as covering maps exactly the asserted maps of Theorem 3.*

Proof. Let $^{N_2}\{p,q\}_{n0}$ be a map arising by application of the algorithm to ρ. As $|\rho(\sigma_1\sigma_2)| = 2$ and $|\rho(\sigma_2\sigma_3)| = N_2$, we have $p = 2m$ and $q = kN_2$. Let be $m = 1$. Then necessarily $k = 1$ or $k = 2$. But $k = 1$ is impossible, as $^{N_2}\{2, N_2\}$ would have two vertices. The case $k = 1$ gives the semihosohedra. If $m = 2$, then necessarily $k = 1, 2$ or 4. The number of vertices of $^2\{4, kN_2\}$ would then be 4, 2, or 1. But such regular maps do not exist ([13] and Theorem 4). Now let be $m > 2$. If we choose 1, σ_3, $\sigma_3\sigma_2$, $\sigma_3\sigma_2\sigma_3$, $\cdots$, $(\sigma_3\sigma_2)^{(N_2-1)/2}$ as a Schreier-transversal, we get the following

Reidemeister-generators of H:

$$\begin{aligned}
u_1 &= \sigma_1, \quad u_2 = \sigma_1^{\sigma_3}, \quad u_3 = \sigma_1^{\sigma_2\sigma_3}, \quad u_4 = \sigma_1^{\sigma_2\sigma_3\sigma_2}, \cdots, \\
u_{N_2} &= \sigma_1^{(\sigma_2\sigma_3)^{(N_2-1)/2}}, \quad u_{N_2+1} = \sigma_2, \quad u_{N_2+2} = \sigma_3^{(\sigma_2\sigma_3)^{(N_2-1)/2}}
\end{aligned} \tag{3.5}$$

and the following defining relations:

$$\begin{aligned}
u_1^2 = u_3^2 = u_{N_2}^2 &= \cdots = u_{N_2+1}^2 = u_{N_2+2}^2 = 1 \\
(u_1 u_{N_2+1})^{2m} = (u_1 u_3)^m = (u_3 u_5)^m &= \cdots = (u_{N_2-2} u_{N_2})^m = 1 \\
(u_{N_2+1} u_{N_2+2})^k &= (u_{N_2} u_{N_2+2})^2 = 1.
\end{aligned} \tag{3.6}$$

The exponents occurring in (3.6) are the exact orders of the respective elements in πH. So

$$\pi H = \langle \pi u_1, \pi u_3, \cdots, \pi u_{N_2}, \pi u_{N_2+1}, \pi u_{N_2+2} \rangle = \langle a, b \rangle \simeq D_{2m}, \tag{3.7}$$

where $a = \pi u_1$ and $b = \pi u_{N_2+1}$. There are exactly 3 conjugacy classes $\mathfrak{C}_i$ $(i = 1, 2, 3)$ of involutions in πH. Let be $\mathfrak{C}_1 := \{(ab)^m\}$. The symmetries of regular dihedra show that we can assume $a \in \mathfrak{C}_1$ and $b_{N_2+1} \in \mathfrak{C}_3$. So by (3.7)

$$\pi u_3 = W_3(a, b), \quad \pi u_5 = W_5(a, b), \cdots, \pi u_{N_2} = W_{N_2}(a, b), \quad \pi u_{N_2+2} = W_{N_2+2}(a, b), \tag{3.8}$$

where the $W_i(a, b)$ denote words in a and b. The relations (3.6) and (3.8) form a presentation of πH.

Let us assume that $\pi u_{N_2+2} \notin \mathfrak{C}_1$. Then πu_{N_2+2} reverses the orientation of the dihedron, therefore

$$a \pi u_{N_2+2} = (ab)^\epsilon. \tag{3.9}$$

As $|a \pi u_3| = m > 2$, we have $\pi u_3 \in \mathfrak{C}_1$, $\pi u_5 \in \mathfrak{C}_1$, $\cdots$, $\pi u_{N_2} \in \mathfrak{C}_1$. So

$$b \pi u_3 = (ab)^{\epsilon_3}, \quad b \pi u_5 = (ab)^{\epsilon_5}, \cdots, b \pi u_{N_2} = (ab)^{\epsilon_{N_2}}. \tag{3.10}$$

(3.9) and (3.10) are the relations (3.8) we looked for. Writing the relators (3.9), (3.10) by (3.5) as corresponding words in the σ_i, we get a system S of relators such that $|[p, q]/\mathrm{Ncl}_{[p,q]} S| \leq 2N_2 m$. But all the relators of S do have even length which is not allowed. Therefore $\pi u_{N_2+2} \in \mathfrak{C}_1$. So $\pi u_{N_2+2} = (ab)^m$ and

$$(\pi\sigma_1\pi\sigma_2)^m \pi\sigma_3^{(\pi\sigma_2\pi\sigma_3)^{(N_2-1)/2}} = 1. \tag{3.11}$$

(3.7) is compatible with (3.4) only for odd m. Write $m = 2n + 1$. As πu_{N_2+2} belongs to the center of πH, whereas b does not, we get $k = 2$. The normal closure in $[4n + 2, 2N_2]$ of the relator corresponding to (3.11) defines a non-orientable regular ${}^{N_2}\{4n + 2, 2N_2\}$ for every n, because it is a torsion-free normal subgroup of index $4N_2(2n + 1)$. The number of edges is $N_2(2n + 1)$, and the characteristic is given by (1.1). The simplest map of the series is drawn below (see the figure in the next page).

By using our method, Scherwa has given the classification for the case $N_2 = 4$.

The non-orientable regular map $^3(6,6)_{no,-6}$
as part of the regular tessellation $(6,6)^*$

Theorem 4: *Let be* $\tau := \sigma_1\sigma_2\sigma_3\sigma_2 \cdot \sigma_3 \cdot \sigma_1\sigma_2\sigma_3\sigma_2$. *Then the regular maps* $^4\{p,q\}$ *are the following ones:* $^4\{3n,4\}_{n0,4-3n}$ *defined by* $\mathrm{Ncl}\langle\tau\rangle$, $^4\{6n,4\}_{n0,4-6n}$ *defined by* $\mathrm{Ncl}\langle(\sigma_1\sigma_2)^{3n}\tau\rangle$, $^4\{12n,8\}_{n0,4-18n}$ *defined by* $\mathrm{Ncl}\langle(\sigma_1\sigma_2)^{3n}\tau\rangle$, $^4\{12n,8\}_{n0,4-18n}$ *defined by* $\mathrm{Ncl}\langle(\sigma_1\sigma_2)^{-3n}\tau\rangle$.

Arguing in the spirit of [10, §4], we get

Theorem 5: *Let* $\{p,q\}_{n0,\chi}$ *be given such that*
     (i) p, q odd
or   (ii) $p \equiv 2 \bmod 4$, q odd prime
or   (iii) $p = 4n$, $q = 3$ and there exists no $\{4n, 2n\}_{n0,\chi}$
Then $G_{\{p,q\}_{n0,\chi}}$ *is not solvable.*

Proof. (i) As $[p,q]^{(ab)} \simeq C_2$, a solvable G would have $G^{ab} \simeq C_2$, so N would be contained in $[p,q]^+$.

(ii) If G were solvable, $[4n+2,q]^{ab} \simeq C_2 \times C_2$ and the condition (3.2) implies that the commutator series of G starts with $|G:G'| = 2$, $|G':G''| = q$, $|G'':G'''| = 2$. (Use Reidemeister-Schreier to get the possible normal subgroups in $[4n+2,q]$ which correspond to G', G'', G'''.) But there exists no 3-step metacyclic group.

(iii) is proved in a similar way. Observe that $[4n,2n]$ is a subgroup of index 3 in $[4n,3]$.

The *Petrie map* $\{p,q\}_r$ is defined by the group

$$G^{p,q,r} := \langle A,B,C : A^p = B^q = C^r = (AB)^2 = (BC)^2 = (CA)^2 = (ABC)^2 = 1\rangle$$
$$\simeq [p,q]/\mathrm{Ncl}\langle(\sigma_1\sigma_2\sigma_3)^r\rangle.$$

(Write $\sigma_1 = BC$, $\sigma_2 = BCA$, $\sigma_3 = CA$. See [5; p.112]). If r is odd, then $G^{p,q,r}$ is a factor group of

$$(2,p,q;r) := \langle R,S \; ; \; R^p = S^q = (RS)^2 = (RSR^{-1}S^{-1})^r = 1\rangle.$$

$(2,p,q;r)$ is a subgroup of index 2 in $G^{p,q,2r}$.

We showed in Remark 2 that for $\chi = -6$ there exists exactly one non-orientable regular map (and its dual), namely the Petrie map $\{7,3\}_9$. If χ decreases, this happens the next time for $\chi = -13$: In fact, $\chi \leq 0$ requires $p,q \geq 3$, hence $3N_2 \leq pN_2 = 2N_1$. This leads to $\chi - N_0 + N_1 = N_2 \leq 2N_1/3$ and $N_1 \leq 3(N_0 - \chi)$. So $q = 2N_1/N_0 \leq 6(1 - \chi/N_0)$, and observing that the non-existence of non-orientable regular maps for $N_0 = 1,2$ and the symmetrie in p and q, we have $p, q \leq 2(3 - \chi)$. The diophantic equation (1.1) has with $3 \leq q \leq p \leq 2(3 - \chi)$ exactly 16 solutions. Theorems 3, 4, 5 discard all cases except, $^{18}\{6,6\}$, $^{13}\{8,4\}$, $^{26}\{6,4\}$, $^{52}\{5,4\}$, $^{78}\{7,3\}$. The orders of the groups in the first four cases show that the 13-Sylow-subgroup is normal in the group of the map, and this is impossible, as there don't exist the corresponding factor maps.

So we are left with $^{78}\{7,3\}$, and G not solvable of order 1092. Hence $G \simeq \mathrm{PSL}(2,13)$, which is also isomorphic to the rotation group of the induced orientable $\{7,3\}_{0,-26}$. There are exactly $\varphi(7)/2 = 3$ regular $\{7,3\}_{0,-26}^{(i)}$. ($i = 1,2,3$) with rotation group isomorphic to $\mathrm{PSL}(2,13)$ [11: Folgerung, p.208].

These maps are defined by $(2,3,7;6)$, $(2,3,7;7)$ and $G^{3,7,13}$ [5; p.96]. So $\{7,3\}^{(3)}_{0,-26}$ is canonically induced by exactly one [Lemma 3] regular $\{7,3\}_{n0,-13}$, namely by $\{7,3\}_{13}$.

Faithful permutation representations of the rotation groups of $\{7,3\}^{(1)}_{0,-13}$ and $\{7,3\}^{(2)}_{0,-13}$ are

$$
\begin{aligned}
R &= (1\ 6\ 9\ 4\ 10\ 3\ 11)(2\ 13\ 14\ 7\ 5\ 12\ 8) \\
S &= (1\ 11\ 9)(3\ 5\ 4)(7\ 12\ 10)(8\ 14\ 13) \qquad \text{and} \\
R^* &= (1\ 11\ 2\ 13\ 6\ 12\ 10)(3\ 14\ 9\ 4\ 5\ 7\ 8) \\
S^* &= (1\ 10\ 4)(3\ 6\ 14)(5\ 12\ 8)(9\ 13\ 11)
\end{aligned}
$$

The permutation groups $\langle R, S \rangle$ and $\langle R^*, S^* \rangle$ are subgroups of the alternating group A_{14}. But the mappings $R \longrightarrow R^{-1}$, $S \longrightarrow S^{-1}$ and $R^* \longrightarrow R^{*-1}$, $S^* \longrightarrow S^{*-1}$ are achieved only by the odd involutions

$$(1\ 13)(2\ 6)(3\ 7)(4\ 12)(5\ 10)(8\ 9)(11\ 14) \quad \text{and} \quad (1\ 3)(2\ 7)(4\ 6)(5\ 13)(8\ 11)(9\ 12)(10\ 14).$$

So by Lemma 3 the map $\{7,3\}_{13}$ and its dual are the only non-orientable regular maps for $\chi = -13$.

4. The crystallographic aspect.

If N is the normal subgroup of $[p,q]$ which defines a non-orientable regular map of genus $h > 1$, then by Theorem 1 $N^{ab} \simeq \mathbf{Z}_2 \times \mathbf{Z}^{h-1}$. The torsion part of N^{ab} gives rise to a normal subgroup N^* such that $[N,N] < N^* < [p,q]$ and $N/N^* \simeq \mathbf{Z}^{h-1}$.

By a theorem due to Zassenhaus [25], every group which is an extension of a free abelian group A of rank g by a finite group G such that G operates effectively on A is isomorphic to a Euclidean crystallographic space group of dimension g. G is then isomorphic to a crystallographic point group. So $[p,q]/N^*$ is isomorphic to a crystallographic space group of dimension $h - 1 = 1 - \chi$, iff

$$C_{[p,q]/N^*}(N/N^*) = N/N^*.$$

On the other hand: As $(N^+)^{ab} \simeq \mathbf{Z}^{2g}$ by Theorem 1 in the orientable case, the group of an orientable regular map of genus $g > 0$ might act as a $2g$-dimensional crystallographic point group.

Let us return to the non-orientable case. As N, $N^+ := N \cap [p,q]^+$ are normal in $[p,q]$ and as $|N : N^+| = 2$, the subgroup $H := \langle [a,b]; a, b \in N \setminus N^+ \rangle$ is normal in $[p,q]$. Moreover, $H \subset [N,N]$ and $N = \langle a \ ; \ a \in N \setminus N^+ \rangle$. Hence we have $H = [N,N]$. Similarly, we define $H_k := \langle a^k : a \in N \setminus N^+, k \in \mathbf{N} \rangle$. H is normal in $[p,q]$. In [1] we show in a more general setting that for even k there exists a normal subgroup N^{**} such that

$$H_k < N^{**} < [p,q] \quad \text{and} \quad N^+/N^{**} \simeq \mathbf{Z}^g \tag{4.1}$$

hold (g the genus of the induced orientable map). So G operates on g-dimensional lattices over the rational integers. This leads to g-dimensional matrix-representations of the group of the map over $\mathbf{Z}$.

Example 1: $N_1 := \mathrm{Ncl}_{[6,4]}\langle(\sigma_1\sigma_2\sigma_3\sigma_2)^2\sigma_3\rangle$ and $N_2 := \mathrm{Ncl}_{[6,4]}\langle(\sigma_1\sigma_2\sigma_3)^3\rangle$ define the two non-orientable regular maps which exist for $p \geq q$ and $\chi = -2$. Both induce the same $\{6,4\}_{o,-4}$. Let $U = \langle\sigma,\tau\rangle$, where $\sigma := \sigma_2\sigma_1\sigma_2\sigma_1\sigma_3$, $\tau := \sigma_1\sigma_2\sigma_3$. We have $\|[p,q]:U\| = 8$ and $U := \langle\sigma,\tau \; ; \; (\sigma\tau)^6 = (\sigma\tau^{-1})^2 = 1\rangle$. Furthermore N_i $(i=1,2)$ is presented by

$$N_i = \langle w_i, x_i, y_i, z_i \; ; \; y_i^{-1}z_iw_ix_i^{-1}z_iy_ix_iw_i = 1\rangle,$$

where $w_1 = \sigma\tau^2$, $x_1 = \tau\sigma\tau$, $y_1 = \tau^2\sigma$, $z_1 = \tau^3\sigma\tau^{-1}$, $w_2 = \tau\sigma^2$, $x_2 = \sigma\tau\sigma$, $y_2 = \sigma^2\tau$, $z_2 = \sigma^3\tau\sigma^{-1}$. The generators w_i, x_i, y_i, z_i do have odd length, hence they are not contained in $[p,q]^+$. Interchanging σ and τ changes N_1 to N_2. In order to write the expressions more simple, we drop the indices of w, x, y, z and N.

Let $\varphi : [p,q] \longrightarrow [p,q]/H_2$ be the natural epimorphism. We have $N^+ = \langle zw, zx, zy, zw^{-1}, zx^{-1}, zy^{-1}, z^2\rangle$ and

$$\varphi(zw) = \varphi(zw^{-1}) =: t_1, \quad \varphi(zx) = \varphi(zx^{-1}) =: t_2, \quad \varphi(zy) = \varphi(zy^{-1}) =: t_3, \quad \varphi(z^2) = 1.$$

t_1, t_2, t_3 form a basis of the free abelian group of rank 3. $[p,q]$ operates on N_1 and N_2 according to

t	t^{σ_1}	t^{σ_2}	t^{σ_3}		t	t^{σ_1}	t^{σ_2}	t^{σ_3}
w	y	x	w^{-1}		w	$y^{-1}z^{-1}y$	$xw^{-1}x^{-1}$	$wy^{-1}z$
x	x^{-1}	w	x	and	x	$wx^{-1}y$	x^{-1}	$y^{-1}xw^{-1}$
y	w	$z^{-1}yw^{-1}$	y^{-1}		y	$z^{-1}yw^{-1}$	$wx^{-1}y$	$y^{-1}zy$
z	$wy^{-1}z$	$wx^{-1}z$	$y^{-1}z^{-1}y$		z	$(w^{-1})^{y^{-1}z}$	$(z^{-1})^{xw^{-1}}$	$y^{x^{-1}y}$

respectively. Using the basis $\{t_1, t_2, t_3\}$, we get for the first map the representation ρ_1 of $[p,q]/N_1^+$:

$$\rho_1(\sigma_1) = \begin{bmatrix} -1 & 0 & 2 \\ -1 & 1 & 1 \\ 0 & 0 & 1 \end{bmatrix}, \quad \rho_1(\sigma_2) = \begin{bmatrix} -1 & 2 & 0 \\ 0 & 1 & 0 \\ 0 & 1 & -1 \end{bmatrix}, \quad \rho_1(\sigma_3) = \begin{bmatrix} 1 & 0 & -2 \\ 0 & 1 & -2 \\ 0 & 0 & -1 \end{bmatrix}$$

and for the second map

$$\rho_2(\sigma_1) = \begin{bmatrix} -1 & 0 & 4 \\ 0 & -1 & 3 \\ 0 & 0 & 1 \end{bmatrix}, \quad \rho_2(\sigma_2) = \begin{bmatrix} -3 & 4 & 0 \\ -2 & 3 & 0 \\ -1 & 1 & 1 \end{bmatrix}, \quad \rho_2(\sigma_3) = \begin{bmatrix} 1 & 0 & -4 \\ 1 & -1 & -2 \\ 0 & 0 & -1 \end{bmatrix}$$

We have $\mathrm{Ker}\rho_1 = \{(\sigma_1\sigma_2\sigma_3)^3, 1\}$, $\mathrm{Ker}\rho_2 = \{(\sigma_1\sigma_2\sigma_3\sigma_2)^2\sigma_3, 1\}$.

Example 2: Let us take the map ${}^3\{6,6\}_{n0,-3}$ (See Theorem 3 and fig.). Take $U = \langle\sigma,\tau,\theta\rangle$, where $\sigma := \sigma_1\sigma_2\sigma_3(\sigma_1\sigma_2)^2$, $\tau := \sigma_2\sigma_1\sigma_3(\sigma_2\sigma_1)^2$, $\theta := (\sigma_1\sigma_2)^2\sigma_1\sigma_2\sigma_3$. Then $\|[6,6]:U\| = 12$ and

$$U = \langle\sigma,\tau,\theta \; ; \; (\sigma\theta)^3 = (\tau\theta)^3 = (\theta\sigma)^3 = 1\rangle.$$

The normal subgroup which defines the map is presented by

$$N := \langle r,s,t,u,v \; ; \; ut^{-1}su^{-1}v^{-1}rt^{-1}vr^{-1}s^{-1} = 1\rangle,$$

where $r := \tau\theta^2$, $s := \theta\tau\theta$, $t := \theta^3$, $u := \sigma\theta^2$, $v := \theta\sigma\theta$. r, s, t, u, v are products of odd length of generators σ_1, σ_2, σ_3, hence elements of $N \setminus N^+$. Rewriting the presentation by $c_0 := t$, $c_1 := u$, $c_2 := v$, $c_3 := r$, $c_4 := s$, we get

$$N = \langle c_i \ (i = 1, \cdots, 4) \ ; \ c_1 c_0^{-1} c_4 c_1^{-1} c_2^{-1} c_3 c_0^{-1} c_2 c_3^{-1} c_4^{-1} = 1 \rangle$$

(The presentation is not quite the normal presentation of Theorem 1). Writing $a_i := c_i c_0^{-1}$, $b_i := c_0 c_i$, we get

$$N^+ = \langle a_i, b_i \ (i = 0, 1, \cdots, 4) \ ; \ b_2 b_3^{-1} a_4^{-1} a_1 a_4 a_1^{-1} b_2^{-1} b_3 = b_4 b_1^{-1} a_2^{-1} a_3 a_2 a_3^{-1} b_4^{-1} b_1 \rangle.$$

[6,6] operates on N according to

$$r^{\sigma_1} = u, \ \ s^{\sigma_1} = v, \ \ t^{\sigma_1} = t, \ \ u^{\sigma_1} = r, \ \ v^{\sigma_1} = s$$

$$r^{\sigma_2} = s^{-1} t r^{-1}, \ s^{\sigma_2} = s \ , t^{\sigma_2} = s r s^{-1} t r^{-1}, \ u^{\sigma_2} = v u s^{-1} t r^{-1}, \ v^{\sigma_2} = u t^{-1} s u^{-1} v^{-1}$$

$$r^{\sigma_3} = s^{-1}, s^{\sigma_3} = r^{-1}, t^{\sigma_3} = t^{-1}, u^{\sigma_3} = v^{-1}, v^{\sigma_3} = u^{-1}.$$

We take the images mod H of $t := tu$, $t := tv$, $t := tr$, $t := ts$ as a basis for the 4-dimensional lattice. With this basis we get the following faithful 4-dimensional representation of $[p, q]/N^+$:

$$\sigma_1 \mapsto \begin{bmatrix} 0 & 0 & 1 & 0 \\ 0 & 0 & 0 & 1 \\ 1 & 0 & 0 & 0 \\ 0 & 1 & 0 & 0 \end{bmatrix}, \quad \sigma_2 \mapsto \begin{bmatrix} -1 & 1 & 1 & -1 \\ 0 & 1 & 0 & -1 \\ 0 & 0 & 1 & -1 \\ 0 & 0 & 0 & -1 \end{bmatrix}, \quad \sigma_3 \mapsto \begin{bmatrix} 0 & 1 & 0 & 0 \\ 1 & 0 & 0 & 0 \\ 0 & 0 & 0 & 1 \\ 0 & 0 & 1 & 0 \end{bmatrix},$$

An Application: The groups H_k can be used to prove certain groups to be infinite. For instance, the groups $(2, 3, 7; r)$ are known to be trivial for $r = 1, 2, 3, 5$ and of order 168, 1092, 1092, 10752 for $r = 4, 6, 7, 8$, respectively. C.C. Sims [20] showed that $(2, 3, 7; 9)$ is infinite, and J. Leech published two notes on $(2, 3, 7; 9)$ [16], [17]. For $r > 9$ nothing seems to be known (see also [4]).

We get Sims' result as an immediate consequence of (4.1). The existence of the regular map $\{7, 3\}_9$ of characteristic -6 implies that H_2 is of infinite index in $[p, q]$. As $(\sigma_1 \sigma_2 \sigma_3)^{18} \in H_2$, $(2, 3, 7; 9)$ is infinite. In the same way, the existence of $\{7, 3\}_{13}$ of characteristic -13 implies that $(2, 3, 7; 13)$ is infinite. For a more general and systematic treatment of the method see [1], where further applications are given.

References

[1]   P. Bergau and D. Garbe, 'Kleinian surfaces and proving groups infinite', to appear.

[2]   H.R. Brahana, 'Regular maps and their groups', *Amer. J. Math.* 49 (1927), 268-284.

[3]   H.S.M. Coxeter, 'Configurations and maps', *Rep. Math. Colloq.* (2) 8(1948), 18-38.

[4] H.S.M. Coxeter, 'The abstract group $G^{3,7,16}$', *Proc. Edinb. Math. Soc.* (2) 13(1962), 47-61.

[5] H.S.M. Coxeter and W.O.J. Moser, *Generators and relators for discrete groups*, 4th ed. Berlin 1980.

[6] A. Dress, 'On the classification and generation of two- and higher-dimensional regular patterns', *Match* 9(1980), 73-80.

[7] W.V. Dyck, 'Gruppentheoretische Studien', *Math. Ann.* 20(1882), 1-45.

[8] V.A. Efremovic, 'Regular polyhedra', (in Russian), *Dokl. Akad. Nauk SSR* 57(1947), 223-226.

[9] R. Fricke and F. Klein, 'Vorlesungen über die Theorie der automorphen Funktionen', Band I. Leipzig (1897).

[10] D. Garbe, 'Ueber die regulaeren Zerlegungen geschlossener orientierbarer Flaechen', *J. reine angew Math.* 237(1969), 39-55.

[11] D. Garbe, 'Ueber eine Klasse von arithmetisch definierbaren Normalteilern der Modulgruppe', *Math. Ann.* 235(1978), 195-215.

[12] D. Garbe, 'A remark on non-symmetric Riemann surfaces', *Arch. d. Math.* 30(1978), 435-437.

[13] A.S. Grek, 'Regular polyhedra of simplest hyperbolic type', (in Russian), *Ivano. Gos. Ped. Inst. Ucen. Zap.* 34(1963), 27-30.

[14] A.S. Grek, 'Regular polyhedra on surfaces of Euler characteristic $\chi = -4$', (in Russian), *Soobsc. Akad. Nauk SSR* 42(1966), 11-15.

[15] A.S. Grek, 'Regular polyhedra on a closed surface whose Euler characteristic is $\chi = -3$'. *AMS Transl.* 78(1968), 127-131.

[16] J. Leech, 'Generators for certain normal subgroups of (2,3,7).' *Proc. Camb. Phil. Soc.* 61(1965), 321-332.

[17] J. Leech, 'Note on the abstract group (2,3,7;9)', *Proc. Camb. Phil. Soc.* 62(1966), 7-10.

[18] J. Scherwa, *Regulaere Karten geschlossener nichtorientierbarer Flaechen*, Diplom-Thesis, Bielefeld 1985.

[19] F.A. Sherk, 'The regular maps on a surface of genus three', *Canad. J. Math.* 11(1959), 452-480.

[20] C.C. Sims, 'On the group (2,3,7;9)', *Notices Amer. Math. Soc.* 11(1964), 687-688.

[21] W. Threlfall, 'Gruppenbilder', *Abh. Saechs. Akad. Wiss. Math.-Phys. Kl.* 41(1932), 1-59.

[22] H.C. Wilkie, 'On non-Euclidean crystallographic groups', *Math.Zeitschr.* 91(1966), 87-102.

[23] J.M. Wills, *Polyhedra in the style of Leonardo, Dali and Escher*, M.C. Escher: Art a. Science. ed. H.S.M. Coxeter et al. Amsterd. 1986.

[24] S. Wilson, 'Riemann surfaces over regular maps', *Can. J. Math.* 30(1978). 763-782.

[25] H. Zassenhaus, 'Ueber einen Algorithmus zur Bestimmung der Raumgruppen', *Comment. Math. Helvet.* 21(1948), 117-141.

Fakultät für Mathematik
Universität Bielefeld
4800 Bielefeld 1
Federal Republic of Germany

PROCEEDINGS OF 'GROUPS – KOREA 1988'

PUSAN, August 1988

THE GROUPS OF AUTOMORPHISMS OF NON–ORIENTABLE HYPERELLIPTIC KLEIN SURFACES WITHOUT BOUNDARY

Emilio Bujalance*, J.A. Bujalance*, G. Gromadzki** and E. Martinez*

1. Introduction.

Let X be a Klein surface (i.e. a compact surface equipped with a dianalytic structure [1]). If X is an orientable surface without boundary then this surface is a classical Riemann surface. A Klein surface X is said to be hyperelliptic if X admits an involution ϕ such that $X/\langle \phi \rangle$ has algebraic genus 0.

The study of automorphism groups of hyperelliptic Riemann surfaces is a classical topic from the 19th century, whilst the studies of groups of automorphisms of hyperelliptic Klein surfaces not being Riemann surfaces have been started to investigate in the recent decade. This paper concerns the problem of the groups of automorphisms of non–orientable hyperelliptic Klein surfaces without boundary. We here solve it completely when the topological genus g of the surface is odd. In the case of bordered hyperelliptic Klein surface, similar results, but concerning only the order of the group, have been established recently in [6].

By the well known correspondence between Klein surfaces and algebraic curves [1], our results can be expressed in terms of the automorphism groups of purely imaginary algebraic curves.

2. NEC groups and hyperelliptic surfaces.

A non–Euclidean crystallographic group (N.E.C. group) Γ is a discrete subgroup of isometries of the hyperbolic plane $\mathcal{D}$ with compact quotient $\mathcal{D}/\Gamma$. Each NEC group Γ has associated with a signature σ that has the form [11]

$$\sigma : (g; \pm; [m_1, \cdots, m_r], \{C_1, \cdots, C_k\}) \tag{2.1}$$

where $C_i = (n_{i1}, \cdots, n_{is_i})$; C_i are called cycle-periods, n_{ij} periods of cycle-periods and m_i proper periods.

The numbers in σ are non–negative integers; m_i and n_{ij} are greater than or equal to 2, and the number g is the topological genus of the surface $\mathcal{D}/\Gamma$ (the algebraic genus $p = \alpha g + k - 1$, where $\alpha = 2$ or $\alpha = 1$ according to the sign '+' or '–' in σ). This surface is orientable or not, according as the sign in σ is '+' or '–' respectively.

If $r = 0$ or $k = 0$, we write in σ [–] or $\{-\}$ respectively. If the number s_i is zero for some i, we denote C_i by $(-)$.

Partially supported by *CAICYT, **Ministerio de Educación y Ciencia de España.

The signature σ determines a canonical presentation of the group Γ, which as shown in [11] and [14], is given by the generators

$$
\begin{array}{ll}
x_i & i = 1, \cdots, r \\
e_i & i = 1, \cdots, k \\
c_{ij} & i = 1, \cdots, k, \quad j = 0, \cdots, s_i \\
a_i, b_i & i = 1, \cdots, g \quad \text{(if sign '+')} \\
d_i & i = 1, \cdots, g \quad \text{(if sign '–')}
\end{array}
$$

subject to the relations

$$
\begin{array}{ll}
x_i^{m_i} = 1 & i = 1, \cdots, r \\
c_{ij-1}^2 = c_{ij}^2 = (c_{ij-1}c_{ij})^{n_{ij}} = 1 & i = 1, \cdots, k; \quad j = 1, \cdots, s_i \\
e_i^{-1} c_{i0} e_i c_{is_i} = 1 & i = 1, \cdots, k
\end{array}
$$

$$
\prod_{i=1}^{r} x_i \prod_{i=1}^{k} e_i \prod_{i=1}^{g} [*] = 1
$$

where $[*]$ is $a_i b_i a_i^{-1} b_i^{-1}$ or d_i^2 according to the sign in σ.

The area of Γ is

$$
\mu(\Gamma) = 2\pi[\alpha g + k - 2 + \sum_{i=1}^{r}(1 - 1/m_i) + 1/2 \sum_{i=1}^{k} \sum_{j=1}^{s_i}(1 - 1/n_{ij})] \tag{2.2}
$$

If Γ is a subgroup of Γ' of index N then the following relation between areas holds:

$$
\mu(\Gamma) = N\mu(\Gamma'). \tag{2.3}
$$

A non–orientable Klein surface without boundary X of algebraic genus $p \geq 2$ can be expressed as $X = \mathcal{D}/\Gamma$ where Γ is an NEC group with signature

$$
(g; -; [-], \{-\}) \tag{2.4}
$$

and if $G = \mathrm{Aut}(X)$ is the full group of automorphisms of X then G is isomorphic to $N_{\mathcal{G}}(\Gamma)/\Gamma$ where $\mathcal{G}$ is the group of all isometries of $\mathcal{D}$ (see [12]).

In [9] the non–orientable hyperelliptic Klein surfaces without boundary are characterized by means of NEC groups. We summarize the results in the following.

Theorem 2.1. *Let $X = \mathcal{D}/\Gamma$ be a non–orientable Klein surface without boundary of genus $g \geq 3$. Then*

(a) *X is hyperelliptic if and only if there exists a unique NEC group Γ_1 containing Γ as a subgroup of index 2, Γ_1 having one of the following signatures*

(i). $(0; \pm; [\overbrace{2, \cdots, 2}^{g}], \{(-)\})$ (ii). $(1; -; [\overbrace{2, \cdots, 2}^{g}], \{-\})$ *(g even)*

(b) *The automorphism of hyperellipticity ϕ where $\langle \phi \rangle = \Gamma_1/\Gamma$ is a central element in the full group of automorphisms of G of X.*

In our paper a hyperelliptic Klein surface (in short HKS) will mean a non–orientable hyperelliptic Klein surface without boundary.

3. Groups of automorphisms of non–orientable hyperelliptic Klein surfaces without boundary.

In this section, we will give a necessary and sufficient conditions for a finite group to be the group of automorphisms of a HKS when a group of hyperellipticity Γ_1 has signature (i) (see theorem 2.1). Thus our results will hold for all HKS for which a group of hyperellipticity has signature (i) and so in particular they are complete for HKS of odd topological genus g.

Lemma 3.1. (1) *Let* $G/\mathbf{Z}_2 \cong \mathbf{Z}_N$ *and* $\mathbf{Z}_N \subseteq G$. *Then*

(i) *if* N *is odd then* $G \cong \mathbf{Z}_{2N}$ *and*

(ii) *if* N *is even then* $G \cong \mathbf{Z}_2 \times \mathbf{Z}_N$ *or* $G \cong \mathbf{Z}_{2N}$.

(2) *Let* $G/\mathbf{Z}_2 \cong \mathbf{D}_{N/2}$ *and* $\mathbf{D}_{N/2} \subseteq G$. *Then*

(i) *if* $N/2$ *is odd then* $G \cong \mathbf{D}_N$ *and*

(ii) *if* $N/2$ *is even then* $G \cong \mathbf{D}_N$ *or* $G \cong \mathbf{Z}_2 \times \mathbf{D}_{N/2}$, *or* $G = \mathbf{U}_{N/2}$, *where* $G = \mathbf{U}_{N/2}$ *is the group with presentation* $\langle x, y \mid x^N, y^2, yxyx^{N/2+1} \rangle$.

Proof. Note first that $\mathbf{Z}_2$ is a central subgroup of G and the second cohomology $\mathrm{H}^2(\mathbf{Z}_N, \mathbf{Z}_2)$ is trivial (if N is odd) or is the cyclic group of order 2 (if N is even). Since $\mathrm{H}^2(G, A)$ classifies the extensions of G by A, the first case follows.

Now let $G/\mathbf{Z}_2 \cong \mathbf{D}_{N/2}$. If $N/2$ is odd then $\mathrm{H}^2(\mathbf{D}_{N/2}, \mathbf{Z}_2) = \mathbf{Z}_2$ and it turns out that there are two such groups: $G = \mathbf{D}_N$ and $G = G_{N/2}$, where $G_{N/2}$ is the group with the presentation $\langle x, y \mid x^N, x^{N/2}y^2, y^{-1}xyx \rangle$ (see [2]). Let H be a subgroup of index 2 in $G_{N/2}$. We will show that H is a cyclic group generated by x. In fact, $y2 \in H$ and so do $x^{N/2}$. Since $x^2 \in H$ and $N/2$ is odd we obtain that $x \in H$, and so H is a cyclic group of order N as required. Hereby $G_{N/2}$ does not contain $\mathbf{D}_{N/2}$

Now let $N/2$ be even. Although in this case $\mathrm{H}^2(\mathbf{D}_{N/2}, \mathbf{Z}_2)$ is a group of order 8, it turns out that there are only six groups G for which $G/\mathbf{Z}_2 \cong \mathbf{D}_{N/2}$ and the complete list of them can be found in [2]. Obviously $\mathbf{D}_N$ and $\mathbf{Z}_2 \times \mathbf{D}_{N/2}$ contain $\mathbf{D}_{N/2}$ as a subgroup. Moreover it is easy to check that (y, x^2) generate the dihedral subgroups of order N in $\mathbf{U}_{N/2}$. Now looking at the possible generators of quotients of G of order 2 and using the Reidemeister–Schreier algorithm for determining the presentation of a subgroup of a given group one can argue that there is no dihedral group $\mathbf{D}_{N/2}$ among subgroups of index 2 in the remaining three groups.

Lemma 3.2. *Let* Γ_1 *be an NEC group with signature* $(0; +; [\overbrace{2, \cdots, 2}^{g}], \{(-)\})$ *and let* Γ^* *be an NEC group containing* Γ_1 *as a normal subgroup of index* N. *Then* Γ^* *has one of the following signatures:*

(i) $(0; +; [N, \overbrace{2, \cdots, 2}^{k}], \{(-)\})$, *where* $k = g/N$,

(ii) $(0; +; [2N, \overbrace{2, \cdots, 2}^{k}], \{(-)\})$, *where* $k = (g-1)/N$,

(iii) $(0;+;[\overbrace{2,\cdots,2}^{h}],\{(N/2,\overbrace{2,\cdots,2}^{l})\})$, *for some h, where* $l = 2g/N - 2h + 2$,

(iv) $(0;+;[\overbrace{2,\cdots,2}^{h}],\{(N,\overbrace{2,\cdots,2}^{l})\})$, *for some h, where* $l = 2(g-1)/N - 2h + 2$.

Proof. Let Γ^* has (2.1) and let p_i be the smallest integer for which $x_i^{p_i} \in \Gamma_1$. Since the only elements of Γ_1 of finite order are those of order 2, we have that $m_i/p_i = 2$ or $m_i/p_i = 1$. For notational convenience assume that $m_i/p_i = 2$ for $i = 1,\cdots,n$ and $m_i/p_i = 1$ for $i = n+1,\cdots,r$. The elliptic elements in the second case do not produce proper periods in Γ_1, whilst in the first case they produce $\sum_{i=1}^{n} N/p_i$ proper periods, all of them being equal to 2 (see [3] and [4]).

Now consider the periods in Γ_1 provided by reflections of Γ^*. Let q_{ij} be the smallest integers for which $(c_{ij}c_{ij-1})^{q_{ij}} \in \Gamma_1$. The same argument, as in the first case, shows that $n_{ij}/q_{ij} = 2$ or 1. Let $E = \{(i,j) \mid n_{ij}/q_{ij} = 2\}$. Clearly the only proper periods in Γ_1 induced by reflections of Γ^* are those provided by the pairs of reflections corresponding to the elements of E, and each such pair produces $N/2q_{ij}$ periods, all of them being equal to 2 (see [3]). As a result

$$g = \sum_{i=1}^{n} N/p_i + \sum_{(i,j)\in E} N/2q_{ij} \tag{3.1}$$

Using (2.3) we obtain

$$\begin{aligned}
g/2 - 1 = {} & N[\alpha g^* - 2 + k + \sum_{i=1}^{n}(1 - 1/m_i) + \sum_{i=n+1}^{r}(1 - 1/m_i) \\
& + 1/2 \sum_{(i,j)\in E}(1 - 1/n_{ij}) + 1/2 \sum_{(i,j)\notin E}(1 - 1/n_{ij})]
\end{aligned} \tag{3.2}$$

The number n has been chosen in such a way that $m_i = 2p_i$ for $i = 1,\cdots,n$ and $m_i = p_i$ for $i = n+1,\cdots,r$, whilst $n_{ij} = 2q_{ij}$ for $(i,j) \in E$, and $n_{ij} = q_{ij}$ for $(i,j) \notin E$. So using (3.1) and (3.2) we obtain the following equation.

$$-1 = \alpha N g^* - 2N + kN + N \sum_{i=1}^{n}(1 - 1/p_i) + N \sum_{i=1}^{k}\sum_{j=1}^{s_i}(1 - 1/q_{ij}). \tag{3.3}$$

Clearly k cannot be bigger than 1, otherwise the right hand side of (3.3) would be ≥ 0. So let $k = 1$. Then $g^* = 0$, otherwise it again would be ≥ 0. Thus we see that Γ^* has signature

$$(0,+,[m_1,\cdots,m_r],\{(n_1,\cdots,n_s)\})$$

and so (3.3) becomes

$$N - 1 = N \sum_{i=1}^{r}(1 - 1/p_i) + N/2 \sum_{j=1}^{s}(1 - 1/q_j). \tag{3.4}$$

Consider two cases:

Case 1. The cycle period of Γ^* is empty. Then $N - 1 = N \sum_{i=1}^{r}(1 - 1/p_i)$ and it is easy to check that the only solution of this equation is that all p_i but one, say p_1, are equal to 1 and $p_1 = N$. Now for $i \neq 1$, $p_i = 1$, and so $m_i = 2$. If $m_1/p_1 = 1$ then $m_1 = N$, whilst if $m_1/p_1 = 2$, $m_1 = N/2$. So Γ^* has a signature (i) or (ii) (the number k can be found using (2.3)).

<u>Case 2</u>. The cycle period is non-empty. Then by [7], [8] two consecutive periods, say n_{s-1}, n_s are equal to 2. Moreover if c_{s-2}, c_{s-1}, c_s are the corresponding reflections, we can assume that $c_{s-1} \in \Gamma_1$, and c_{s-2}, $c_s \notin \Gamma_1$ (see [8]). Thus q_{s-1}, $q_s = 2$ and so the second summand of the right hand side of (3.4) is $\geq N/2$. Now all p_i are equal to 1, otherwise the first summand of the right hand side is also $\geq N/2$ and is $\geq N$, a contradiction. The relation (3.4) becomes

$$N/2 - 1 = N/2 \sum_{j=1}^{s-2} (1 - 1/q_j)$$

and the only solution of it is that all q_j but one, say q_1, are equal to 1 and $q_1 = N/2$. Now since all $p_i = 1$, $m_i = 2$ for $i = 1, \cdots, r$. For $q_j = 1$, the corresponding periods are equal to 2, and for $q_1 = N/2$ the corresponding period is $N/2$ or N according as $n_1/q_1 = 1$ or 2. So in this case Γ^* has a signature (iii) or (iv) (the number l can be found as in the first case using (2.3)).

Remark 3.3. From the proof of the previous lemma follows not only that Γ^* has one of the specified signatures but also an information how the group Γ_1 sits in Γ^*.

In the case of signatures (i), (ii) all elliptic elements x_i but x_1 belong to Γ_1 whilst $x_1^N \in \Gamma_1$ and N is the smallest integer with this property.

In the case of signatures (iii), (iv) all elliptic element x_i must belong to Γ_1, all c_i, but c_1, do not belong to Γ_1 and $c_i c_{i+1} \in \Gamma_1$ for $i = 1, \cdots, l - 2$, whilst $(c_0 c_1)^{N/2} \in \Gamma_1$ and $N/2$ is the smallest integer with this property.

Lemma 3.4. *Let Γ_1 and Γ^* be NEC groups as in the previous lemma and assume that Γ is an NEC group with signature (2.4) being a subgroup of Γ_1 of index 2 and normal in Γ^*. Let $G = \Gamma^*/\Gamma$. Then*

(1) *If Γ^* has signature* (i) *then $G = Z_{2N}$ or $G = Z_2 \times Z_N$.*

(2) *If Γ^* has signature* (ii) *then $G = Z_{2N}$.*

(3) *If Γ^* has signature* (iii) *then $G = D_N$ or $G = Z_2 \times D_{N/2}$ or $G = U_{N/2}$, where*
   $U_{N/2} = \langle x, y \mid x^N, y^2, yxyx^{N/2+1} \rangle$.

(4) *If Γ^* has signature* (iv) *then $G = D_N$.*

Proof. Let $G^* = \Gamma^*/\Gamma_1$ and $H = \Gamma_1/\Gamma \cong Z_2$. Clearly $G/H \cong G^*$. Let x_i, e_i, c_i be the canonical generators of Γ^*. We will employ the notations introduced in the proof of the previous lemma.

(1) Let Γ^* have signature (i). We show that in this case $p_1 = N$ and so the image of x_1 in G is an element of order N. So $Z_N \subseteq G$. Thus by Lemma 3.1 $G \cong Z_{2N}$ or $G \cong Z_2 \times Z_N$.

(2) Let Γ^* have signature (ii). Then since Γ has no periods, x_1 induces an element of order $2N$ in G. Hereby $G \cong Z_{2N}$.

(3) Now assume that Γ^* has signature (iii). We show that the pair of reflections c_0, c_1 corresponding to the period $N/2$ satisfy c_0, $c_1 \notin \Gamma_1$ and c_0, c_1 induces in G^* an element of order $N/2$. So $G^* \cong D_{N/2}$. Moreover c_0, $c_1 \notin \Gamma$ and $c_0 c_1$ also induces an element of order $N/2$ in G. Thus the result follows from Lemma 3.1.

(4) Finally let Γ^* have a signature (iv) and let c_0, c_1 be the pair of reflections corresponding to

the period N as in the previous case. Then since Γ has no proper periods and no period cycles c_0, $c_1 \notin \Gamma_1$ and their product induces an element of order N in Γ. Therefore $G \cong D_N$.

Theorem 3.5. *Let X be a HKS of odd topological genus g. Then $\mathrm{Aut}(X)$ may be one of the following groups Z_{2N}, $Z_2 \times Z_N$, D_N or $Z_2 \times D_{N/2}$. Furthermore*

(i) *There exists a HKS of genus g having Z_{2N} as the group of automorphisms if and only if $N|g - 1$ and $N \neq g - 1$ or $N|g$, $N \neq g$, and N is odd.*

(ii) *There exists a HKS of genus g having $Z_2 \times Z_N$ $(N \neq 2)$ as the group of automorphisms if and only if $N|g$, $N \neq g$ and N is even.*

(iii) *There exists a HKS of genus g having D_N as the group of automorphisms if and only if $N|2(g-1)$ and N is even or $N|2g$, $4\nmid N$.*

(iv) *There exists a HKS of genus g having $Z_2 \times D_{N/2}$ as the group of automorphisms if and only if $N|2g$ and $4|N$.*

Proof. Let X be a HKS of odd topological genus $g \geq 3$. Then $X = \mathcal{D}/\Gamma$, where Γ is an NEC group with signature (2.4). Since X is a HKS there exists an involution $\phi \in \mathrm{Aut}(X)$ such that $X/\langle\phi\rangle$ is a surface of genus 0. By the theorem 2.1 ϕ is a central element in $\mathrm{Aut}(X)$. Let $\langle\phi\rangle = \Gamma_1/\Gamma$ and let G be a group of automorphisms of X containing ϕ. Then $G = \Gamma^*/\Gamma$ for some NEC group Γ^*, containing Γ_1 as a normal subgroup. By Theorem 2.1 Γ_1 has signature $(0; \pm; [\overset{g}{\overbrace{2, \cdots, 2}}], \{(-)\})$. The Lemma 3.2 describes the possible signatures for Γ^* and Lemma 3.4 gives us necessary conditions for a group G to be represented in such case as a quotient Γ^*/Γ. Given a homomorphism $\theta : \Gamma^* \longrightarrow G$ and a central element ϕ of G of order 2 let π be the canonical projection $G \longrightarrow G/\langle\phi\rangle$ and $\theta^* = \phi \circ \theta$. We have to investigate homomorphisms θ from Γ^* onto groups specified in Lemma 3 such that $\mathrm{Ker}\theta = \Gamma$ and $\mathrm{Ker}\theta^* = \Gamma_1$, where Γ is a group with signature (2.4) and Γ_1 is a group with signature $(0; \pm; [\overset{g}{\overbrace{2, \cdots, 2}}], \{(-)\})$.

1. Let Γ^* have signature (i). By Lemma 3.4 $G = \Gamma^*/\Gamma$ may be only Z_{2N} or $Z_2 \times Z_N$. Assume that $G = Z_{2N}$. We have to see whether or not a homomorphism $\theta : \Gamma^* \longrightarrow Z_{2N} = \langle x \mid x^{2N} \rangle$ satisfying the conditions in question exists. Since $\mathrm{Ker}\theta$ is a surface group, θ must preserve the orders of the canonical generators of Γ^*. So θ is forced to be defined as follows $\theta(x_1) = x^2$, $\theta(x_2) = \cdots = \theta(x_{k+1}) = x^N$, $\theta(c) = x^N$ and $\theta(e) = x^{-(kN+2)}$. Clearly $w = x_2 c$ is a non-orientable element in $\mathrm{Ker}\theta$ and so by [10] $\mathrm{Ker}\theta$ is a non-orientable surface group. But clearly this homomorphism is an epimorphism if and only if N is odd. Let this will be the case. Then $\theta^* : \Gamma^* \longrightarrow Z_{2N}/\langle x^N \rangle \cong Z_N = \langle \bar{x} \mid \bar{x}^N \rangle$ is given as follows $\theta^*(x_1) = (\bar{x})^2$, $\theta^*(x_2) = \cdots = \theta^*(x_{k+1}) = 1$, $\theta^*(c) = 1$ and $\theta^*(e) = (\bar{x})^{-2}$ and so $\mathrm{Ker}\theta^* = \Gamma_1$, by the proof of Lemma 3.2 (see Remark 3.3). Moreover the canonical Fuchsian subgroup $(\Gamma^*)^+$ of Γ^* has the signature $(0; \pm; [N, N, \overset{2k}{\overbrace{2, \cdots, 2}}], \{-\})$ (see [13]) and so $k > 1$ (i.e. $N \neq g$) is a maximal signature (see [12]). Hence the signature of Γ^* is also maximal and so Γ^* can be chosen to be a maximal NEC group and so for the surface $X = \mathcal{D}/\Gamma$ just considered, Z_{2N} can be assumed to be the full group of automorphisms. The signature $(0, +, [N, 2], \{-\})$ corresponding to the case $g = N$ is not maximal, and so for $g = N$ the group Z_{2N} cannot be the full group of automorphisms

of a HKS of genus g (see Proposition 2.4 [5]).

Now let $G = Z_2 \times Z_N$ and let N be even. Consider the homomorphism $\theta : \Gamma^* \longrightarrow Z_2 \times Z_N = \langle x, y \mid x^2, y^N, [x,y] \rangle$ defined by $\theta(x_1) = xy$, $\theta(x_2) = \cdots = \theta(x_{k+1}) = x$, $\theta(c) = x$ and $\theta(e) = y^{-1}x^{(k+1)}$. Since θ preserves the orders of the canonical generators of Γ^* and $w = x_2c$ is a non-orientable element in $\mathrm{Ker}\theta$, $\mathrm{Ker}\theta = \Gamma$. Moreover for the homomorphism $\theta^* : \Gamma^* \longrightarrow Z_2 \times Z_N/\langle x \rangle = \langle y \mid y^N \rangle$ we have $\theta^*(x_1) = y$, $\theta^*(x_2) = \cdots = \theta^*(x_{k+1}) = 1$, $\theta^*(c) = 1$ and $\theta^*(e) = y^{-1}$. So by the proof of Lemma 3.2 $\mathrm{Ker}\theta^* = \Gamma_1$.

As in the case of Z_{2N} we can argue that for $N \neq g$ a dianalytic structure on X can be so chosen that Z_{2N} is the full group of its automorphisms whilst this is not the case for $N = g$.

2. Let Γ^* have signature (ii). By Lemma 3.4 $G = \Gamma^*/\Gamma$ may be only the cyclic group of order $2N$. We will show that this is so. Let $\theta : \Gamma^* \longrightarrow Z_{2N} = \langle x \mid x^{2N} \rangle$ be the homomorphism defined by $\theta(x_1) = x$, $\theta(x_2) = \cdots = \theta(x_{k+1}) = x^N$, $\theta(c) = x^N$ and $\theta(e) = x^{-(kN+1)}$. Clearly $\mathrm{Im}(\theta) = Z_{2N}$. Moreover x_2c is a non-orientable element in $\mathrm{Ker}\theta$ and so $\mathrm{Ker}\theta = \Gamma$. Now for the homomorphism $\theta^* : \Gamma^* \longrightarrow Z_{2N}/\langle x^N \rangle \cong Z_N = \langle \bar{x} \mid \bar{x}^N \rangle$ we have $\theta^*(x_1) = \bar{x}$, $\theta^*(x_2) = \cdots = \theta^*(x_{k+1}) = 1$, $\theta^*(c) = 1$ and $\theta^*(e) = (\bar{x})^{-1}$. So by Remark 3.3 $\mathrm{Ker}\theta^* = \Gamma_1$. As in the previous case we argue that for $N \neq g - 1$, Z_{2N} can be assumed to be the full group of automorphisms of the surface just constructed, whereas this is not the case for $N = g - 1$.

3. Let Γ^* be the group with signature (iii). By Lemma 3.4 $G = \Gamma^*/\Gamma$ may be only one of the groups D_N, $Z_2 \times D_{N/2}$ or $U_{N/2}$. Clearly N must be even in this case, otherwise D_N would have the trivial center whilst the remaining two groups are not defined. We will show first that the group D_N is a group of automorphisms of a surface in question if and only if $N/2$ is odd. Let $\theta : \Gamma^* \longrightarrow D_N = \langle x, y \mid x^2, y^2, (xy)^N \rangle$ be the homomorphism we are looking for. Let z be the central element of order 2 in D_N for which $D_N/\langle z \rangle \cong D_{N/2}$. Then by Remark 3.3 $\theta(x_1), \cdots, \theta(x_h)$ must be equal to z, $\theta(c_0) = \theta(c_{l+1}) = az^{\varepsilon_0}$ ($\varepsilon_0 = 0$ or 1), $\theta(c_i) = bz^{\varepsilon_i}$ ($\varepsilon_i = 0$ or 1) for $i = 1, \cdots, l-1$, $\theta(c_l) = z$ for some elements a, b of order 2 whose product has order $N/2$. So in particular $D_N \cong Z_2 \times D_{N/2}$. But the last is the case if and only if $N/2$ is odd, i.e. $4 \nmid N$.

Conversely let $N/2$ be odd. Let $h = 0$ and let $\theta : \Gamma^* \longrightarrow D_N$ be the homomorphism defined by $\theta(c_0) = \theta(c_{l+1}) = y$ $\theta(c_i) = xyx$ for $1 \leq i \leq l-1$ and i odd, $\theta(c_i) = xyx(xy)^{N/2}$ for $1 \leq i \leq l-1$ and i even, $\theta(c_l) = (xy)^{N/2}$, and $\theta(e) = 1$. It is easy to check that $\mathrm{Im}(\theta) = D_N$ and $\mathrm{Ker}\theta = \Gamma$ whilst $\mathrm{Ker}\theta^* = \Gamma_1$. The same argument as used before shows that the dianalytic structure on X just considered can be so chosen that $\mathrm{Aut}(X) = D_N$.

Now let $N/2$ be even and let for $h = 0$ $\theta : \Gamma^* \longrightarrow Z_2 \times D_{N/2} = \langle z \mid z^2 \rangle \otimes \langle x, y \mid x^2, y^2, (xy)^{N/2} \rangle$ be the homomorphism defined by $\theta(c_0) = \theta(c_{l+1}) = x$, $\theta(c_i) = y$ for $1 \leq i \leq l-1$ and i odd, $\theta(c_i) = yz$ for $1 \leq i \leq l-1$ and i even, $\theta(c_l) = z$, and $\theta(e) = 1$. As in the previous case we show that $\mathrm{Ker}\theta = \Gamma$ and for the homomorphism $\theta^* : \Gamma^* \longrightarrow Z_2 \times D_{N/2}/\langle z \rangle = D_{N/2}$, $\mathrm{Ker}\theta^* = \Gamma_1$. Again the same argument as used before shows that the dianalytic structure on X just considered can be so chosen that $\mathrm{Aut}(X) = Z_2 \times D_{N/2}$.

So it remains to show that the group $U_{N/2}$ cannot stand for a group of automorphisms of a surface in question. If this were so, then arguing as in the case of the dihedral group, one can show that the corresponding homomorphism were considered as the one defined there. Consequently, $U_{N/2}$ must be isomorphic to the direct product $Z_2 \times D_{N/2}$.

4. Now consider the group Γ^* with signature (iv). By the Lemma 3.4 $G = \Gamma^*/\Gamma$ may be only the dihedral group D_N of order $2N$. Clearly N is even since D_N has trivial center for N odd. Let for $h = 0$ $\theta : \Gamma^* \longrightarrow D_N = \langle x, y \mid x^2, y^2, (xy)^N \rangle$ be the homomorphism defined by $\theta(c_0) = \theta(c_{l+1}) = y$, $\theta(c_1) = x$, $\theta(c_i) = (xy)^{N/2}x$ for $2 \leq i \leq l-1$ and i even, $\theta(c_i) = x$ for $2 \leq i \leq l-1$ and i odd, $\theta(c_l) = (xy)^{N/2}$, and $\theta(e) = 1$. Clearly $\mathrm{Im}\theta = D_N$ and $(c_0c_1)^{N/2}c_1$ is a non-orientable element in $\mathrm{Ker}\theta$. So $\mathrm{Ker}\theta = \Gamma$. Moreover for the homomorphism $\theta^* : \Gamma^* \longrightarrow D_N/\langle(xy)^{N/2}\rangle \cong D_{N/2} = \langle \bar{x}, \bar{y} \mid \bar{x}^2, \bar{y}^2, (\bar{x}\bar{y})^{N/2} \rangle$ we have $\theta^*(c_0) = \theta^*(c_{l+1}) = \bar{y}$, $\theta^*(c_1) = \cdots = \theta^*(c_{l-1}) = \bar{x}$, $\theta^*(c_l) = 1$, $\theta^*(e) = 1$. By the proof of the Lemma 3.2 (see Remark 3.3) $\mathrm{Ker}\theta^* = \Gamma_1$. As before we argue that the dianalytic structure on the surface just considered can be chosen in such a way that D_N is the full group of its automorphisms.

References

[1]   N.L. Alling and N. Greenleaf, 'Foundations of the theory of Klein surfaces', Lecture Notes in Math. 219, Springer, Berlin, 1971.

[2]   R. Brandt and H. Stichtenoth, 'Die automorphismengruppen hyperelliptischer Kurven', *Manuscripta Math.* 55(1986), 83–92.

[3]   E. Bujalance, 'Proper periods of normal NEC subgroups with even index', *Rev. Mat. Hisp.-Amer.* (4) 41(1981), 121–127.

[4]   E. Bujalance, 'Normal subgroups of NEC groups', *Math. Z.* 178(1981), 331–341

[5]   E. Bujalance, 'Automorphism groups of compact planar Klein surfaces', *Manuscripta Math.* 56(1986), 105–124.

[6]   E. Bujalance, J.A. Bujalance and E. Martinez, 'On the automorphism group of hyperelliptic Klein surfaces' *Michigan Math. J.*, to appear.

[7]   E. Bujalance and E. Martinez, ' A remark on NEC groups of surfaces with boundary', *Bull. London Math. Soc.*, to appear.

[8]   J.A. Bujalance, 'Normal subgroups of even index of an NEC group', *Archiv Math.* 49(1987), 470–478.

[9]   J.A. Bujalance, 'Q–Hyperelliptic compact non–orientable Klein surfaces without boundary', *Kodai Math. J.*, to appear.

[10]   A.M.H. Hoare and D. Singerman, 'The orientability of subgroups of plane groups', London Math. Soc. Lecture Notes Series, 71(1982), 221–227.

[11]   A.M. Macbeath, 'The classification of non–Euclidean crystallographic groups', *Canad. J. Math.* 6(1967), 1192–1205.

[12]   D. Singerman, 'Finitely maximal Fuchsian groups', *J. London Math. Soc.* (2) 6(1972), 29–38.

[13] D. Singerman, 'On the structure of non–Euclidean crystallographic groups', *Proc. Cambridge Phil. Soc.* 76(1974), 233–240.

[14] H.C. Wilkie, 'On non–Euclidian crystallographic groups', *Math. Z.* 91(1966), 87–102.

Additional Refernces added in proof.

C. Maclachlan, 'Smooth coverings of hyperelliptic surfaces', *Quart. J. Math.* II 22(1971), 117–123.

D. Singerman, ' Symmetries and pseudosymmetries of hyperelliptic surfaces', *Glasgow Math. J.* 21(1980), 39–49.

(*)   Departamento de Matemáticas Fund.
       Facultad de Ciencias
       UNED
       28040 – Madrid
       Spain

(**) Instytut Matematyki WSP
       Chodkiewicza 30
       85–064 Bydgoszcz
       Poland

ON CERTAIN ONE-RELATOR PRODUCTS OF CYCLIC GROUPS

Colin M. Campbell, E.F. Robertson and R.M. Thomas*

1. Introduction.

A *one-relator product* of cyclic groups $G_1, G_2, \cdots, G_m$ is the quotient $(G_1 * G_2 * \cdots * G_m)/N$, where N is the normal closure of a cyclically reduced word in $G_1 * G_2 * \cdots * G_m$. In this paper, we are concerned with one-relator products of a cyclic group of order 2 with a cyclic group of order n, sometimes known as $(2, n)$-*groups*, i.e, with groups defined by presentations of the form

$$\langle a, \ b : a^2 = b^n = w(a, b) = 1 \rangle,$$

where $w(a, b)$ is some word in the generators a and b, and also with the associated groups of deficiency zero defined by presentations of the form

$$\langle a, \ b : a^2 = 1, \ b^n = w(a, b) \rangle,$$

or, equivalently,

$$\langle a, \ b : a^2 = w_1(a, b) = 1 \rangle,$$

where $w_1(a, b) = w(a, b)b^{-n}$.

We are particularly interested in determining which groups defined by presentations of this form are finite. Clearly, if the deficiency zero group is finite and of known structure, the structure of the corresponding $(2, n)$-group is readily determined.

In some cases, the structure of such groups has been determined. For example, groups defined by presentations of the form

$$\langle a, \ b : a^2 = ab^h ab^i = 1 \rangle$$

are easily seen to be either finite and metacyclic, or else infinite. The structures of groups defined by presentations of the form

$$\langle a, \ b : a^2 = ab^h ab^i ab^j = 1 \rangle$$

are largely known by the results in [2, 4, 5, 6, 7]. In [11], it was determined which of the groups defined by the presentations

$$\langle a, \ b : a^2 = b^3 = w(a, b) = 1 \rangle,$$

where $w(a, b)$ has length less than 24, are finite. The groups $G(r, n)$ defined by the presentations

$$\langle a, \ b : a^2 = b^n = abab^r ab^{-r} ab^{-1} = 1 \rangle$$

were shown in [14] to be infinite if $d > 3$, or if $d = 3$ and n is even, where $d = (r+1, n)$. Further results on the groups $G(r, n)$, and the associated group of deficiency zero, may be found in [8]. In [1], it was shown that a group defined by a presentation of the form

$$\langle a, b : a^m = b^n = w(a, b)^r = 1 \rangle,$$

where m, n, $r > 1$, is infinite if $1/m + 1/n + 1/r \leq 1$.

For convenience, we shall let $G(n; i(1), i(2), \cdots, i(r))$ denote the group defined by the presentation

$$\langle a, b : a^2 = b^n = ab^{i(1)}ab^{i(2)} \cdots ab^{i(r)} = 1 \rangle$$

and $H(n; i(1), i(2), \cdots, i(r))$ the group defined by

$$\langle a, b : a^2 = 1, \ b^n = ab^{i(1)}ab^{i(2)} \cdots ab^{i(r)} \rangle.$$

The structure of the groups $G(n; h, i, j, k)$ with $h + i + j + k = 0$ and $h, i, j, k \in \{\pm 1, \pm 2\}$ was determined in [10]. In this paper, we shall consider the groups $H(n; h, i, j, k)$ under this constraint. Clearly, if $n = h + i + j + k = 0$, the group $G(n; h, i, j, k)$ is infinite, since the abelianized quotient is isomorphic to $C_2 \times C$, so we shall assume that $n \neq 0$. In [9] it was shown that the group defined by the presentation

$$\langle a, b : a^2 = 1, \ ab^2ab^{-2}ab^{-1}ab = b^n \rangle$$

is a finite soluble group of order $|(2n(n+2)g_n f_{(n,3)}|$ for n odd, isomorphic to SD_{16} for $n = 2$, metabelian of order 240 for $n = 4$ and order 80 for $n = -4$, and infinite otherwise. So we shall concentrate on the remaining cases here. Our results are:

Theorem A. *The group $H(n; 2, -2, 2, -2)$ is:*
 (i)   *isomorphic to D_{18} for $n = 1$,*
 (ii) *isomorphic to D_{14} for $n = -1$,*
 (iii) *soluble of order 528 and derived length 3 for $n = 3$,*
 (iv) *soluble of order 240 and derived length 3 for $n = -3$,*
 (v)   *infinite for n even or $|n| > 3$.*

Theorem B. *The groups $H(n; 2, 2, -2, -2)$ and $H(n; 2, 1, -2, -1)$ are infinite for n even, and metacyclic of order $2|n|^3$ for n odd.*

Theorem C.
 (i) $H(1; 1, -1, 1, -1)$ *is isomorphic to D_{10}.*
 (ii) $H(-1; 1, -1, 1, -1)$ *is isomorphic to D_6.*
 (iii) $H(2; 1, -1, 1, -1)$ *is metabelian of order 48.*
 (iv) $H(-2; 1, -1, 1, -1)$ *is isomorphic to SD_{16}.*
 (v) $H(3; 1, -1, 1, -1)$ *is soluble of order 336 and derived length 3.*
 (vi) $H(-3; 1, -1, 1, -1)$ *is soluble of order 48 and derived length 3.*

(vii) $H(n; 1, -1, 1, -1)$ *is infinite for* $|n| > 3$.

Theorem D. *The group* $H(n; 1, 2, -2, -1)$ *is metabelian of order*

$$2 \cdot |n||n + 6| \cdot (2^{|n|} - (-1)^{|n|})/3.$$

The notation used in this paper is reasonably standard. We use $|\cdot|$ to denote either the order of a group or the modulus of an integer, the context, hopefully, making it clear as to which is intended. If G and H are groups, we let $G \times H$ denote the direct product, and $G * H$ the free product, of G and H. For any group G, G' denotes the commutator subgroup of G. We let C_n, D_n, Q_n, E_n and SD_n denote the cyclic, dihedral, generalized quaternion, elementary abelian and semidihedral groups respectively of order n, C denoting the infinite cyclic group, and A_n the alternating group of degree n. We let (f_n) and (g_n) denote the *Fibonacci* and *Lucas* sequences of numbers defined inductively by

$$f_0 = 0, \quad f_1 = 1, \quad f_{n+2} = f_n + f_{n+1}$$
$$g_0 = 2, \quad g_1 = 1, \quad g_{n+2} = g_n + g_{n+1}.$$

Lastly, if a and b are integers, we let (a, b) denote their highest common factor.

2. First reductions.

In this section, we prove some preliminary results, pointing out certain isomorphisms between the groups under consideration. We first have:

Proposition 2.1. *The group* $H(n; h, i, j, k)$ *is isomorphic to the group* $H(-n; -h, -i, -j, -k)$.

Proof. This follows immediately from replacing the generator b by b^{-1}.

Proposition 2.2. *The group* $H(n; h, i, j, k)$ *is isomorphic to the group* $H(n; j, i, h, k)$.

Proof. The relation $ab^h ab^i ab^j ab^k = b^n$ is equivalent to $b^{-k} ab^{-j} ab^{-i} ab^{-h} a = b^{-n}$, and hence to $ab^{-j} ab^{-i} ab^{-h} ab^{-k} = b^{-n}$. Now replacing b by b^{-1} yields the result.

In view of Propositions 2.1 and 2.2, we may assume that either $h = 1$, or else $h = 2$ and $|j| = 2$. We deal with the latter case first, in Section 3 for $h = 2$, $j = 2$, and in Section 4 for $h = 2$, $j = -2$.

This will leave the case $h = 1$. Since the group defined by the presentation

$$\langle b, c : bc^i = c^{n+i}b, \ cb^i = b^{n+i}c \rangle$$

is infinite if $(i, n - i) > 1$, i.e., if $(i, n) > 1$, and metacyclic of order $|n|^3$ if $(i, n - i) = 1$, by the Theorem in [3], we have:

Proposition 2.3. $H(n; 1, i, -1, -i)$ *is:*

(i)   metabelian of order $2|n|^3$ for $i = 1$ or -1,

(ii)   metabelian of order $2|n|^3$ for $i = 2$ or -2 and n odd,

(iii) infinite for $i = 2$ or -2 and n even.

Proof. This follows immediately from the Theorem in [3], since we have a normal subgroup of index 2 in $H(n; 1, i, -1, -i)$ with presentation

$$\langle b, c : bc^i b^{-1} c^{-i} = c^n, \ cb^i c^{-1} b^{-i} = b^n \rangle,$$

where $c = aba$, which is metabelian of order $|n|^3$, or infinite, according as $(i, n) = 1$, or $(i, n) > 1$, respectively.

Given this result, the following cases remain to be considered:

$$H(n; 1, -1, 1, -1); \qquad H(n; 1, -2, 2, -1); \qquad H(n; 1, -1, 2, -2)$$
$$H(n; 1, -1, -2, 2); \qquad H(n; 1, 2, -2, -1);$$

The third case need not be considered separately, in view of the following result:

Proposition 2.4. *The group $H(n; 1, -1, 2, -2)$ is isomorphic to the group $H(n; 1, -2, 2, -1)$.*

Proof. Consider the presentation

$$\langle a, b : a^2 = 1, \ abab^{-1}ab^2ab^{-2} = b^n \rangle.$$

Since $b^{-2}abab^{-1}ab^2 a = b^n$, we have $b^n a = (b^{-1}ab^2)^{-1}a(b^{-1}ab^2)$, and so $(b^n a)^2 = 1$, i.e., $ab^n a = b^{-n}$. Thus $bab^{-1}ab^2ab^{-2}a = b^{-n}$, and hence $ab^{-1}ab^2ab^{-2}ab = b^{-n}$. So we have

$$\langle a, b : a^2 = 1, \ abab^{-1}ab^2ab^{-2} = b^n, \ ab^{-1}ab^2ab^{-2}ab = b^{-n} \rangle.$$

Conversely, the third relation gives that $ab^{-n} = (b^{-2}ab)^{-1}a(b^{-2}ab)$, so that $ab^{-n}a = b^n$. Then we have that $bab^{-1}ab^2ab^{-2}a = b^{-n}$, and we may deduce the second relation, which is therefore redundant. So we have

$$\langle a, b : a^2 = 1, \ ab^{-1}ab^2ab^{-2}ab = b^{-n} \rangle.$$

Replacing b by b^{-1} yields the result.

A similar result disposes of the need to consider the fourth case separately:

Proposition 2.5. *The group $H(n; 1, -1, -2, 2)$ is isomorphic to the group $H(n; 1, 2, -2, -1)$.*

In fact, these groups are isomorphic to the groups studied in [9]:

Proposition 2.6. *The group $H(n; 1, 2, -2, -1)$ is isomorphic to the group $H(-n; 2, -2, -1, 1)$.*

Proof. Consider the presentation

$$\langle a, b : a^2 = 1, abab^2ab^{-2}ab^{-1} = b^n \rangle.$$

As in the proofs of Propositions 2.5 and 2.6, we get that $ab^n a = b^{-n}$, and then $b^2ab^{-2}ab^{-1}ab = b^{-n}$. So we have

$$\langle a, b : a^2 = 1, abab^2ab^{-2}ab^{-1} = b^n, ab^2ab^{-2}ab^{-1}ab = b^{-n} \rangle.$$

Since the last relation gives that $ab^{-n}a = b^n$ as usual, the second relation is redundant, and the result follows.

So we have reduced this part of the problem down to two cases. We deal with the groups $H(n; 1, -1, 1, -1)$ in Section 5 and the groups $H(n; 1, -2, 2, -1)$ in Section 6.

3. The groups H(n;2,i,2,k).

In this section, we deal with the groups $H(n; h, i, j, k)$ with $h = 2$ and $j = 2$. Since $h+i+j+k = 0$, we must have $i = k = -2$, and so we have the presentation

$$\langle a, b : a^2 = 1, ab^2ab^{-2}ab^2ab^{-2} = b^n \rangle.$$

This has homomorphic image $G = G(n; 2, -2, 2, -2)$ with presentation

$$\langle a, b : a^2 = b^n = ab^2ab^{-2}ab^2ab^{-2} = 1 \rangle,$$

and G is an infinite group for n even by Proposition 2.1 (i) of [10]. So we may assume that n is odd, in which case G is isomorphic to $G(n; 1, -1, 1, -1)$ with presentation

$$\langle a, b : a^2 = b^n = abab^{-1}abab^{-1} = 1 \rangle$$

by Proposition 2.1 (ii) of [10], and $G(n; 1, -1, 1, -1)$ is infinite for $|n| > 3$ by Theorem 4.1 (iii) of [10]. So we have Theorem A (v) and the only remaining presentations to consider here are

$$\langle a, b : a^2 = 1, ab^2ab^{-2}ab^2ab^{-2} = b \text{ or } b^{-1} \rangle,$$

and

$$\langle a, b : a^2 = 1, ab^2ab^{-2}ab^2ab^{-2} = b^3 \text{ or } b^{-3} \rangle.$$

In the first case, let $c = aba$ and N be the normal subgroup $\langle b, c \rangle$ of index 2. Then N has presentation

$$\langle b, c : c^2b^{-2}c^2b^{-2} = b \text{ or } b^{-1}, b^2c^{-2}b^2c^{-2} = c \text{ or } c^{-1} \rangle.$$

Since $(c^2b^{-2}c^2b^{-2})^{-1} = b^2c^{-2}b^2c^{-2}$, we have that $b = c^{-1}$. So N has presentation

$$\langle b : b^{-8} = b \text{ or } b^{-1} \rangle,$$

and we see that N is isomorphic to C_9 or C_7. Since $aba = c = b^{-1}$, $H(1; 2, -2, 2, -2)$ is isomorphic to D_{18} and $H(-1; 2, -2, 2, -2)$ to D_{14}, thus proving Theorem A (i) and (ii).

In the second case, we let $c = aba$ and $N = \langle b, c \rangle$ again, and we have

$$\langle b, c : c^2 b^{-2} c^2 b^{-2} = b^3 \text{ or } b^{-3}, \ b^2 c^{-2} b^2 c^{-2} = c^3 \text{ or } c^{-3} \rangle.$$

Since $(c^2 b^{-2} c^2 b^{-2})^{-1} = b^2 c^{-2} b^2 c^{-2}$, we have that $b^3 = c^{-3} = z$ (say), which is then central in N, and then

$$\langle b, c, z : b^3 = c^{-3} = z, \ c^2 b^{-2} c^2 b^{-2} = z \text{ or } z^{-1}, \ b^2 c^{-2} b^2 c^{-2} = z^{-1} \text{ or } z \rangle.$$

The last relation is redundant via the penultimate, and so we have

$$\langle b, c, z : b^3 = c^{-3} = z, \ c^2 b^{-2} c^2 b^{-2} = z \text{ or } z^{-1} \rangle.$$

Now, since $b^3 = z$, we have $b^{-2} = z^{-1} b$, and, similarly, $c^2 = c^{-1} z^{-1}$. So the third relation is equivalent, via the first two, to $c^{-1} z^{-1} z^{-1} b c^{-1} z^{-1} z^{-1} b = z$ or z^{-1}, and so we have

$$\langle b, c, z : b^3 = c^{-3} = z, \ (c^{-1} b)^2 = z^5 \text{ or } z^3 \rangle.$$

If we replace the generator c by $d = c^{-1}$, we have

$$\langle b, d, z : b^3 = c^3 = z, \ (bd)^2 = z^5 \text{ or } z^3 \rangle.$$

We introduce a new generator $e = d^{-1} b^{-1}$ to get

$$\langle b, d, e, z : b^3 = z, \ d^3 = z, \ e^2 = z^{-5} \text{ or } z^{-3}, \ bde = 1 \rangle.$$

This is the group

$$\begin{matrix} < & 3 & 3 & 2 & > \\ & 1 & 1 & -5 & 0 \end{matrix} \qquad \text{or} \qquad \begin{matrix} < & 3 & 3 & 2 & > \\ & 1 & 1 & -3 & 0 \end{matrix}$$

in the notation of [12], and has order

$$144 \cdot \left| \frac{1}{3} + \frac{1}{3} - \frac{5}{2} - 0 \right| = 264, \quad \text{or} \quad 144 \cdot \left| \frac{1}{3} + \frac{1}{3} - \frac{3}{2} - 0 \right| = 120$$

respectively by (4.15) of [12]. So $H(3; 2, -2, 2, -2)$ has order 523, and $H(-3; 2, -2, 2, -2)$ has order 240.

Let G be the group $H(3; 2, -2, 2, -2)$ or $H(-3; 2, -2, 2, -2)$. We note that, since $N/\langle z \rangle$ is isomorphic to A_4 and $[G : G'] = 6$ in each case $G'/\langle z \rangle$ is isomorphic to E_4. Adding the relation $z^2 = 1$ to those for N yields

$$\langle b, d, z : b^3 = d^3 = (bd)^2 = z, \ z^2 = 1 \rangle,$$

which is a presentation for the binary tetrahedral group $\langle 2, 3, 3 \rangle$ whose Sylow 2-subgroup is isomorphic to Q_8. So N does not have an abelian Sylow 2-subgroup, and hence G' is not abelian. We see, therefore, that G is a soluble group of derived length 3. This completes the proof of Theorem A.

4. The groups H(n;2,i,-2,k).

In this section, we investigate the groups $H(n; 2, i, -2, k)$. Since $h + i + j + k = 0$, we must have $k = -1$. Moreover, $H(n; 2, i, -2, -i)$ is isomorphic to $H(-n; -2, -i, 2, i)$ by Proposition 2.1, and then to $H(-n; 2, -i, -2, i)$ by Proposition 2.2. So we may assume that $i > 0$. Hence we have either

$$\langle a, b : a^2 = 1, ab^2ab^2ab^{-2}ab^{-2} = b^n \rangle,$$

or

$$\langle a, b : a^2 = 1, ab^2abab^{-2}ab^{-1} = b^n \rangle.$$

Since $G(n; 2, 2, -2, -2)$ is infinite for n even by Proposition 2.1 (i) of [10], and $G(n; 2, 1, -2, -1)$ is isomorphic to $G(n; 1, 2, -1, -2)$ by Propositions 2.2 and 2.3 of [10], which is infinite for n even by Theorem 3.2(iii) of [10], we may assume that n is odd in each case, say $n = 2k - 1$.

In the first case, let $c = aba$ and N be the normal subgroup $\langle b, c \rangle$ of index 2. Then N has presentation

$$\langle b, c : c^2b^2c^{-2}b^{-2} = b^n, \ b^2c^2b^{-2}c^{-2} = c^n \rangle.$$

Since $(c^2b^2c^{-2}b^{-2})^{-1} = b^2c^2b^{-2}c^{-2}$, we have that $b^n = c^{-n} = z$ (say), and hence we have

$$\langle b, c, z : b^n = c^{-n} = z, \ c^2b^2c^{-2}b^{-2} = z, \ b^2c^2b^{-2}c^{-2} = z^{-1} \rangle.$$

The last relation is redundant, so that we have

$$\langle b, c, z : b^n = c^{-n} = z, \ c^2b^2c^{-2}b^{-2} = z \rangle.$$

Since $c^n = z^{-1}$, we have that $c^{2k} = cz^{-1}$. Then $c^2b^2c^{-2} = b^2z$ implies that $c^{2k}b^2c^{-2k} = b^2z^k$, and hence that $cb^2c^{-1} = b^2z^k$. This, in turn, gives that $c^2b^2c^{-2} = b^2z^{2k}$, and hence that $z^{2k} = z$, so that $z^n = 1$. Hence we have

$$\langle b, c, z : b^n = c^{-n} = z, \ z^n = 1, \ cb^2c^{-1} = b^2z^k, \ c^2b^2c^{-2} = b^2z \rangle.$$

The last relation is now redundant. Since $cb^2c^{-1} = b^2z^k$, we have

$$cbc^{-1} = c(b^{2k}z^{-1})c^{-1} = (cb^2c^{-1})^kz^{-1} = b^{2k}z^{m-1} = bz^m$$

where $m = k^2$. So we have

$$\langle b, c, z : b^n = c^{-n} = z, \ z^n = 1, \ cbc^{-1} = bz^m, \ cb^2c^{-1} = b^2z^k \rangle.$$

Since $2m \equiv k \pmod{n}$, the last relation is redundant, and we have

$$\langle b, c, z : b^n = c^{-n} = z, \ z^n = 1, \ cbc^{-1} = bz^k \rangle.$$

This is a presentation for metacyclic group of order $|n|^3$, and so we have proved the first part of Theorem B.

We now turn our attention to the group $H(n; 2, 1, -2, -1)$ with presentation

$$\langle a, b : a^2 = 1, ab^2abab^{-2}ab^{-1} = b^n \rangle.$$

If we let $a_1 = bab^{-1}$, $a_2 = b^2ab^{-2}$, $\cdots$, $a_{n-1} = b^{n-1}ab^{1-n}$, $z = b^n$, then we have the relations

$$a a_2 a_3 a_1 = z, \quad a_1 a_3 a_4 a_2 = z, \quad \cdots, \quad a_{n-1} a_1 a_2 a = z,$$

and hence

$$a a_2 = z a_1 a_3, \quad a_1 a_3 = z a_2 a_4, \quad \cdots, \quad a_{n-1} a_1 = z a a_2,$$

which yield that $a a_2 = z^n a a_2$, and hence that $z^n = 1$. Now consider the normal subgroup N of index 2 in $H(n; 2, 1, -2, -1)$ with presentation

$$\langle b, c : c^2 b c^{-2} b^{-1} = b^n, \ b^2 c b^{-2} c^{-1} = c^n \rangle.$$

Then the relation $c^2 b c^{-2} = b^{n+1}$ gives that $c^2 b^2 c^{-2} = b^{2n+2}$, and hence that $c^2 b^2 c^{-2} b^{-2} = b^{2n}$. Similarly, $b^2 c^2 b^{-2} c^{-2} = c^{2n}$. Hence $b^{2n} = c^{-2n}$.

Now let $z = b^n$ as above, so that $z^n = 1$. Then $c^{2n} = z^{-2}$, and hence $c^{2kn} = z^{-2k} = z^{-1}$, i.e., $c^{n(n+1)} = z^{-1}$, i.e., $c^n = z^{-1}$. So we have

$$\langle b, c, z : b^n = c^{-n} = z, \ z^n = 1, \ c^2 b c^{-2} = bz, \ b^2 c b^{-2} = cz^{-1} \rangle.$$

The relation $c^2 b c^{-2} = bz$ gives that $c^{2k} b c^{-2k} = bz^k$, i.e., that $cbc^{-1} = bz^k$. In turn, $cbc^{-1} = bz^k$ implies that $c^2 b c^{-2} = bz^{2k} = bz$. So we have

$$\langle b, c, z : b^n = c^{-n} = z, \ z^n = 1, \ cbc^{-1} = bz^k, \ b^2 c b^{-2} = cz^{-1} \rangle.$$

The last relation is redundant, and N is seen to be metacyclic of order $|n|^3$. So we have completed the proof of Theorem B.

5. The groups H(n;1,-1,1,-1).

In this section, We prove Theorem C.

(i) Here we have the presentation

$$\langle a, b : a^2 = 1, \ abab^{-1}abab^{-1} = b \rangle.$$

Since $bab^{-1} \cdot a \cdot (bab^{-1})^{-1} = ab$, we have $(ab)^2 = 1$, and hence $aba = b^{-1}$. So we have

$$\langle a, b : a^2 = b^5 = (ab)^2 = 1 \rangle,$$

which is a presentation for D_{10}.

(ii) Here we have

$$\langle a, b : a^2 = 1, \ abab^{-1}abab^{-1} = b^{-1} \rangle.$$

So we have $abab^{-1}aba = 1$, i.e., $ab^{-1}ab^2 = 1$, i.e., $aba = b^2$. So $b^4 = b$, and we have

$$\langle a, b : a^2 = b^3 = (ab)^2 = 1 \rangle,$$

which is a presentation for D_6.

(iii) In this case, we have
$$\langle a,\ b : a^2 = 1,\ abab^{-1}abab^{-1} = b^2 \rangle.$$

Let $c = aba$, N be the normal subgroup $\langle b,\ c \rangle$ of index 2. Then N has presentation
$$\langle b,\ c : (cb^{-1})^2 = b^2,\ (bc^{-1})^2 = c^2 \rangle.$$

Since $(cb^{-1})^2 = (bc^{-1})^{-2}$, we have $b^2 = c^{-2}$. Also $bc^{-1}b = c^3$, so that $b^{-1}c^{-1}b = b^{-2}c^3 = c^5$. Since $b^2 = c^{-2}$, we have $c^{-1} = b^{-2}c^{-1}b^2 = c^{-25}$, and hence $c^{24} = 1$. Thus we have
$$\langle b,\ c : c^{24} = 1,\ b^2 = c^{-2},\ (cb^{-1})^2 = b^2,\ (bc^{-1})^2 = c \rangle.$$

The last relation is redundant via the previous two, and the third may be rewritten as $b^{-1}cb^{-1} = c^{-1}b^2$, i.e., $b^{-1}cb = c^{-1}b^4 = c^{-5}$. So we have
$$\langle b,\ c : c^{24} = 1,\ b^2 = c^{-2},\ b^{-1}cb = c^{-5} \rangle.$$

Since $b^{-1}cb = c^{-5}$, we have $b^{-1}c^2b = c^{-10}$, and so $c^2 = c^{-10}$, i.e., $c^{12} = 1$. So we have
$$\langle b,\ c : c^{12} = 1,\ b^2 = c^{-2},\ b^{-1}cb = c^{-5} \rangle,$$

which is a presentation for a metacyclic group of order 24. The result now follows.

(iv) Here we have
$$\langle a,\ b : a^2 = 1,\ abab^{-1}abab^{-1} = b^{-2} \rangle,$$
i.e. $\langle a,\ b : a^2 = 1,\ (ab)^3 = ba \rangle$. We introduce a new generator $d = ab$ and delete $b = ad$ to get
$$\langle a,\ d : a^2 = 1,\ d^3 = ada \rangle,$$

Clearly $d^9 = d$, i.e., $d^8 = 1$, and we have a presentation for SD_{16}.

(v) Let $H = H(3; 1, -1, 1, -1)$ with presentation
$$\langle a,\ b : a^2 = 1,\ abab^{-1}abab^{-1} = b^3 \rangle.$$

We let $a_1 = bab^{-1}$, $a_2 = b^2ab^{-2}$, $z = b^3$, and M be the normal subgroup $\langle a,\ a_1,\ a_2,\ z \rangle$ of index 3 in H. We proceed as in Chapter 7 of [13] to derive a presentation for M. Working with the coset representatives $\{1,\ b,\ b^2\}$, we get a, a_1, a_2 and z as generators for M, and relations $a^2 = a_1^2 = a_2^2 = 1$, $(aa_1)^2 = z$, $(a_1a_2)^2 = z$, $(a_2zaz^{-1})^2 = z$. Since $(aa_1)^2 = z$, we have $az = a_1aa_1$, and so $(az)^2 = 1$. We may similarly deduce that $(a_1z)^2 = (a_2z)^2 = 1$, and so we have
$$\langle a,\ a_1,\ a_2,\ z : a^2 = a_1^2 = a_2^2 = (az)^2 = (a_1z)^2 = (a_2z)^2 = 1,$$
$$(aa_1)^2 = z, (a_1a_2)^2 = z,\ (a_2zaz^{-1})^2 = z \rangle$$

Now:
$$a_2zaz^{-1}a_2zaz^{-1} = a_2az^{-2}a_2az^{-2} = a_2aa_2z^2az^{-2} = (a_2a)^2z^{-4},$$

so that the last relation is equivalent to $(a_2a)^2 = z^5$. We introduce $u = aa_1$ and delete $a_1 = au$ to get

$$\langle a,\ u,\ a_2,\ z : a^2 = (au)^2 = a_2^2 = (az)^2 = (auz)^2 = (a_2 z)^2 = 1,$$
$$u^2 = z,\ (aua_2)^2 = z,\ (a_2 a)^2 = z^5\rangle.$$

Now introduce $v = a_2 a$ and delete $a_2 = va$ to get

$$\langle a,\ u,\ v,\ z : a^2 = (au)^2 = (va)^2 = (az)^2 = (auz)^2 = (vaz)^2 = 1,$$
$$u^2 = z,\ (auva)^2 = z,\ v^2 = z^5\rangle.$$

Since $aza = z^{-1}$, the relation $(auva)^2 = z$ may be rewritten as $(uv)^2 = z^{-1}$. The relation $(auz)^2 = 1$ is redundant via $u^2 = z$ and $(au)^2 = (az)^2 = 1$. The relation $(vaz)^2 = 1$ is redundant via $v^2 = z^5$ and $(va)^2 = (az)^2 = 1$. The relation $(az)^2 = 1$ is then redundant via $u^2 = z$ and $(au)^2 = 1$. So we have

$$\langle a,\ u,\ v,\ z : a^2 = (au)^2 = (va)^2 = 1,\ u^2 = z,\ (uv)^2 = z^{-1},\ v^2 = z^5\rangle.$$

Let P denote the normal subgroup $\langle u,\ v,\ z\rangle$ of index 2 in M with coset representatives $\{1,\ a\}$. Working as in Chapter 7 of [13] again, we have generators $u,\ u_1 = aua^{-1},\ v,\ v_1 = ava^{-1},\ z,$ $z_1 = aza^{-1},\ t = a^2$ for P, with relations $t = 1,\ u = u_1^{-1},\ v = v_1^{-1},\ u^2 = z,\ u_1^2 = z_1,\ (uv)^2 = z^{-1},$ $(u_1 v_1)^2 = z_1^{-1},\ v^2 = z^5,\ v^{12} = z^{15}$. Removing the trivial generator t and deleting $u_1,\ v_1$ and z_1 in turn, yields

$$\langle u,\ v,\ z : u^2 = z,\ v^2 = z^5,\ (uv)^2 = z^{-1},\ (u^{-1}v^{-1})^2 = z\rangle.$$

Now:

$$(u^{-1}v^{-1})^2 = (uu^{-2}vv^{-2})^2 = (uz^{-1}vz^{-5})^2 = (uv)^2 z^{-12} = z^{-13},$$

so that the last relation is equivalent to $z^{14} = 1$. So we have

$$\langle u,\ v,\ z : u^2 = z,\ v^2 = z^5,\ (uv)^2 = z^{-1},\ z^{14} = 1\rangle.$$

We may replace the relation $z^{14} = 1$ by $u^{28} = 1$, and the relation $v^2 = z^5$ by $v^2 = u^{10}$. Then $(uv)^2 = v^2 v^{-2} uvuv = z^5 uv^{-1} uv$. So that the relation $(uv)^2 = z^{-1}$ is equivalent to $v^{-1} uv = u^{-1} z^{-6} = u^{-13} = u^{15}$. So we have

$$\langle u,\ v,\ z : u^2 = z,\ u^{28} = 1,\ v^2 = u^{10},\ v^{-1}uv = u^{-15}\rangle.$$

The generator z is redundant, and we see that we have a presentation for a metacyclic group of order 56.

Now $[H : H'] = 6$, and, since $[H : P] = 6$ and P is contained in H' (since $b^3 \in H'$, and $aa_1 = abab^{-1} \in H'$, etc.), $P = H'$. So H is soluble of derived length 3 and order 336 as required.

(vi) Now consider $H = H(-3; 1, -1, 1, -1)$ with presentation

$$\langle a,\ b : a^2 = 1,\ abab^{-1}abab^{-1} = b^{-3}\rangle.$$

A simialar argument to that used in case (v) yields a subgroup P of index 6 in H with presentation

$$\langle u,\ v,\ z : u^2 = z,\ v^2 = z^{-3},\ (uv)^2 = z,\ z^2 = 1\rangle.$$

which is a presentation for the quaternion group Q_8 of order 8. So H is again soluble of derived length 3, but this time has order 48.

(vii) If $|n| > 3$, then $H(n; 1, -1, 1, -1)$ is infinite, since the homomorphic image $G(n; 1, -1, 1, -1)$ is infinite by Theorem 4.1 (iii) of [10].

6. The groups $H(n;1,-2,2,-1)$.

In this section, we consider the groups $H(n; 1, -2, 2, -1)$ with presentation

$$\langle a, b : a^2 = 1, \ abab^{-2}ab^2ab^{-1} = b^n \rangle$$

As usual, let $c = aba$, $N = \langle b, c \rangle$, so that N has presentation

$$\langle b, c : cb^{-2}c^2b^{-1} = b^n, \ bc^{-2}b^2c^{-1} = c^n \rangle.$$

Since $(cb^{-2}c^2b^{-1})^{-1} = bc^{-2}b^2c^{-1}$, we have that $b^n = c^{-n} = z$ (say), and so

$$\langle b, c, z : b^n = c^{-n} = z, \ cb^{-2}c^2b^{-1} = z, \ bc^{-2}b^2c^{-1} = z^{-1} \rangle.$$

The last relation is redundant. We introduce a new generator $x = c^{-1}b$, and then, using the fact that z is central, we may rewrite the relation $cb^{-2}c^2b^{-1} = z$ as $b^{-2}cbb^{-1}cb^{-1}c = z$, i.e., $b^{-1}x^{-1}bx^{-2} = z$, i.e., $b^{-1}x^{-1}b = x^2z$. So we have

$$\langle b, c, x, z : b^n = c^{-n} = z, \ x = c^{-1}b, \ b^{-1}xb = x^{-2}z^{-1} \rangle,$$

and we may delete $c = bx^{-1}$ to get

$$\langle b, x, z : b^n = (bx^{-1})^{-n} = z, \ b^{-1}xb = x^{-2}z^{-1} \rangle.$$

Now

$$\begin{aligned}
b^{-2}xb^2 &= (x^{-2}z^{-1})^{-2}z^{-1} = x^4z, \\
b^{-3}xb^3 &= (x^{-2}z^{-1})^4z = x^{-8}z^{-3}, \\
b^{-4}xb^4 &= (x^{-2}z^{-1})^{-8}z^{-3} = x^{16}z^5,
\end{aligned}$$

and, in general,

$$b^{-m}xb^m = x^pz^q, \tag{$*$}$$

where $p = (-2)^m$, and

$$q = -1 + 2 - 4 + 8 - \cdots + (-1)^m 2^{m-1} = \frac{1}{3}[(-2)^m - 1].$$

We introduce a new generator y defined by

$$y = \begin{cases} z & \text{if } n > 0, \\ z^{-1} & \text{if } n < 0, \end{cases}$$

to get

$$\langle b, x, y, z : b^{|n|} = y, \ (bx^{-1})^{|n|} = y^{-1}, \ b^{-1}xb = x^{-2}z^{-1}, y = z \text{ or } z^{-1} \rangle.$$

Now the relation $(bx^{-1})^{|n|} = y^{-1}$ may be rewritten as $(xb^{-1})^{|n|} = y$, and hence as

$$x \cdot b^{-1} x b \cdot b^{-2} x b^2 \cdot \dots \cdot b^{1-|n|} x b^{|n|-1} \cdot b^{-|n|} = y,$$

i.e., $x^r z^s y^{-1} = y$, where

$$r = 1 - 2 + 4 - 8 + \cdots + (-2)^{|n|-1} = \frac{1}{3}[1 - (-2)^{|n|}]$$

and

$$
\begin{aligned}
s &= -1 + 1 - 3 + 5 - \cdots + \frac{1}{3}[(-2)^{|n|-1} - 1] \\
&= \frac{1}{3}[(-2) + (-2)^2 + \cdots + (-2)^{|n|-1} - (|n| - 1)] \\
&= \frac{1}{3}[\frac{1}{3}[(-2) - (-2)^{|n|}] - (|n| - 1)] \\
&= \frac{1}{9}[1 - (-2)^{|n|} - 3|n|],
\end{aligned}
$$

and hence as $x^r = z^t$, where

$$
t = \begin{cases}
\frac{1}{9}[(-2)^n + 3n + 17] & \text{if } n > 0, \\
\frac{1}{9}[(-2)^{-n} - 3n - 19] & \text{if } n < 0.
\end{cases}
$$

Now (*) and the relation $x = b^{-|n|} x b^{|n|}$ give that $x^{3r} = z^{-r}$, and so $z^{3t} = x^{3r} = z^{-r}$. So $z^u = 1$, where

$$u = 3t + r = \frac{1}{3}[(-2)^n + 3n + 17] + \frac{1}{3}[1 - (-2)^n] = n + 6$$

if $n > 0$, and

$$u = 3t + r = \frac{1}{3}[(-2)^{-n} - 3n - 19] + \frac{1}{3}[1 - (-2)^{-n}] = -(n + 6)$$

if $n < 0$. So we have

$$\langle b, x, z : b^n = z, x^r = z^t, z^u = 1, b^{-1}xb = x^{-2}z^{-1} \rangle.$$

This is a presentation for a metabelian group of order $|nru|$. So we have proved Theorem D.

Acknowledgements

The third author would like to thank Hilary Craig for all her help and encouragement.

References

[1] G. Baumslag, J.W. Morgan and P.B. Shalen, 'Generalized triangle groups', *Math. Proc. Cambridge Phil. Soc.* 102(1987), 25-31.

[2]   C.M. Campbell, H.S.M. Coxeter and E.F. Robertson, 'Some families of finite groups having two generators and two relations', *Proc. Roy. Soc. London* 357A (1977), 423-438.

[3]   C.M. Campbell and E.F. Robertson, 'On a group presentation due to Fox', *Canad. Math. Bull.* 19(1976), 247-248.

[4]   C.M. Campbell and E.F. Robertson, 'Classes of groups related to $F^{a,b,c}$', *Proc. Roy. Soc. Edinburgh* 78A (1978), 209-218.

[5]   C.M. Campbell and E.F. Robertson, 'On 2-generator 2-relation soluble groups', *Proc. Edinburgh Math. Soc.* 23(1980), 269-273.

[6]   C.M. Campbell and E.F. Robertson, 'Groups related to $F^{a,b,c}$ involving Fibonacci numbers', in C. Davis, B. Grunbaum and F.A. Sherk (eds.), *The Geometric Vein* (Springer-Verlag, 1982), 569-576.

[7]   C.M. Campbell and E.F. Robertson, 'On the $F^{a,b,c}$ conjecture', *Mitt. Math. Sem. Giessen* 164(1984), 25-36.

[8]   C.M. Campbell, E.F. Robertson and R.M. Thomas, 'On groups related to Fibonacci groups', in *Proceedings of the Singapore Group Theory Conference*, to appear.

[9]   C.M. Campbell, E.F. Robertson and R.M. Thomas, 'On finite groups of deficiency zero involving the Lucas numbers', to appear.

[10] C.M. Campbell and R.M. Thomas, 'On (2,n)-groups related to Fibonacci groups', *Israel J. Math.* 58(1987), 370-380.

[11] M.D.E. Conder, 'Three-relator quotients of the modular group', *Quart. J. Math.* 38(1987), 427-447.

[12] J.H. Conway, H.S.M. Coxeter and G.C. Shephard, 'The centre of a finitely generated group', *Tensor* 25(1972), 405-418.

[13] D.L. Johnson, *Presentations of Groups*, London Math. Soc. Lecture Notes 22 (Cambridge University Press, 1976).

[14] R.M. Thomas, 'Some infinite Fibonacci groups', *Bull. London Math. Soc.* 15 (1983), 384-386.

Mathematical Institute   * Department of Computing studies
University of St. Andrews   University of Leicester
St. Andrews KY16 9SS   Leicester LE1 7RH
Scotland   England

PROCEEDINGS OF 'GROUPS – KOREA 1988'

PUSAN, August 1988

EFFICIENT PRESENTATIONS FOR FINITE SIMPLE GROUPS AND RELATED GROUPS

Colin M. Campbell, E.F. Robertson and P.D. Williams*

1. Introduction.

Let G be a finite group. A group H of maximal order with the properties that there is a subgroup A with $A \leq Z(H) \cap H'$ and $H/A \cong G$ is called a *covering group* of G. In general H is not unique but A is unique and is called the *Schur multiplier* $M(G)$ of G. For details see [1, 15, 26]. In the case where G is perfect then G has a unique covering group which we denote by $G^{\sim}$.

Schur [22] showed that any presentation for G with n generators requires at least $n+\text{rank}(M(G))$ relations. If G has a presentation with n generators and precisely $n + \text{rank}(M(G))$ relations we say that G is *efficient*. Not all groups are efficient and examples of soluble groups with trivial multiplier which are not efficient were given by Swan [24]. Further details of such groups are given in [1], [25], and [26].

In Section 2 of this paper we describe recent progress in investigating the efficiency of finite simple groups. If G is a finite simple group then $M(G^{\sim})$ is trivial, see [19], [26]. Hence to prove $G^{\sim}$ is efficient we require to find a presentation for $G^{\sim}$ with an equal number of generators and relations. Such presentations are called *balanced*. Notice that a finite group with a balanced presentation is necessarily efficient. The problem of finding balanced presentations for covering groups of finite simple groups is, in general, harder than finding efficient presentations for the simple groups themselves.

Finally in Section 3 we consider the problem of efficient presentations for direct products of simple groups. Given two groups G_1 and G_2 then the Schur-Künneth formula [22] asserts

$$M(G_1 \times G_2) = M(G_1) \times M(G_2) \times (G_1 \otimes G_2).$$

Thus, when G_1 or G_2 is perfect, $M(G_1 \times G_2) = M(G_1) \times M(G_2)$ so the multiplier of a direct product of simple groups is the direct product of multipliers of the simple groups.

2. Finite simple groups and their covering groups.

In this section we review progress on the efficiency of finite simple groups and their covering groups. There are relatively few general results. The first was an efficient presentation for $PSL(2,p)$,

p a prime greater that 3, by Zassenhaus [28]. Prior to this, $PSL(2,5) = A_5$ was the only member of this class which was known to be efficient, essentially shown by Hamilton in 1856. The Zassenhaus result was incomplete and, following an idea by H.S.M. Coxeter, the following theorem was proved by J.G. Sunday.

Theorem [23]. *For p a prime greater than 3, $PSL(2,p)$ has the efficient presentation*

$$PSL(2,p) = \langle a,\ b \mid a^p = 1,\ b^2 = (ab)^3,\ (a^t ba^4 b)^2 = 1 \rangle$$

where $2t \equiv 1 \pmod{p}$.

Using similar methods to Zassenhaus, Campbell and Robertson proved the efficiency of $SL(2,p)$, the covering group of $PSL(2,p)$, by finding the balanced presentation in the following theorem.

Theorem [2]. *For p a prime greater than 3, $SL(2,p)$ has the efficient presentation*

$$SL(2,p) = \langle x,\ y \mid x^2 = (xy)^3,\ (xy^4 xy^t)^2 y^p x^{2k} = 1 \rangle$$

where $t = (p+1)/2$ *and k is the integer part of $p/3$.*

There are other general results which, although they do not give efficient presentations, do come relatively close. General results of this type have helped obtain efficient presentations in particular cases. In particular, work on $PSL(2,q)$, q a prime power, is described in [10] while the case of $PSL(2,p^2)$, p a prime, is described in detail in [27].

For the simple groups of order less than 10^6, excluding the groups $PSL(2,q)$, q a prime power, work on efficient presentations has used as a starting point the minimal permutation generators given by McKay and Young in [20]. Presentations satisfied by these permutation generators are given by Cannon, McKay and Young [11] for those groups G with $|G| < 10^5$ and by Campbell and Robertson [3], in the case $10^5 < |G| < 10^6$. Methods to attempt to make these presentations efficient were first given in [3]. These methods were refined by Kenne [17] and [18] and further developed by Jamali and Robertson [13], [14].

In the table below we list the simple groups of order less than 10^6 excluding $PSL(2,p)$, p a prime. We give the order and Schur multiplier (see [12]) together with references to efficient presentations of both the simple group and its covering group. When the group has trivial multiplier $G = G^*$ and we put a * under the entry for the covering group. The reference Note 1 is to this paper, see the Notes following the table.

TABLE

G	$|G|$	$M(G)$	Efficiency of G	Efficiency of $G\tilde{\ }$
A_6	360	6	[3]	[21]
SL(2,8)	504	1	[2]	*
A_7	2520	6	[3]	[9]
SL(2,16)	4080	1	[5]	*
PSL(3,3)	5616	1	[3] (see also [4])	*
PSU(3,3)	6048	1	[17],[18]	*
PSL(2,25)	7800	2	[5],[21],[27]	[9]
M_{11}	7920	1	[14],[18]	*
PSL(2,27)	9828	2	[21],[27]	[9]
A_8	20160	2	[3]	unknown
PSL(3,4)	20160	3×4^2	[3]	unknown
PSp(4,3)	25920	2	[3]	unknown
Sz(8)	29120	2^2	[3]	[9]
SL(2,32)	32736	1	[9],[18]	*
PSL(2,49)	58800	2	[21],[27]	[9]
PSU(3,4)	62400	1	[3]	*
M_{12}	95040	2	[3]	unknown
PSU(3,5)	126000	3	[14]	unknown
J_1	175560	1	[14]	*
A_9	181440	2	[8]	unknown
SL(2,64)	262080	1	[9]	*
PSL(2,81)	265680	2	[8]	[8]
PSL(3,5)	372000	1	unknown	*
M_{22}	443520	12	[14]	unknown
J_2	604800	2	[6],[8]	[6],[8]
PSL(2,121)	885720	2	[10]	unknown
PSL(2,125)	976500	2	[10]	Note 1
PSp(4,4)	979200	1	unknown	*

TWO LARGER GROUPS

PSL(2,169)	2413320	2	[8]	[8]
PSL(2,361)	23522760	2	[8]	[8]

Note 1. In [10] the efficient presentation

$$PSL(2,5^3) = \langle a,\ b \mid a^2 = b^3 = (ab)^4(ab^{-1})^{14}(ab)^4(ab^{-1})^{-7} = 1\rangle$$

is given. This can be used to find the following efficient presentation for $SL(2,5^3)$.

$$SL(2,5^3) = \langle a,\ b \mid a^2 = b^3,\ (a^{-1}b)^4(ab^{-1})^{14}(a^{-1}b)^4(ab^{-1})^{-7} = 1\rangle.$$

Note 2. The two presentations for J_2 given in [6] and [8] are different in the sense that they are satisfied by distinct minimal permutation generators. The presentation given in [8] is satisfied by the generators 19.21 of [20] while that in [6] is satisfied by 19.15 of [20]. This fact is easily checked with the library of simple groups [7] with the group theory system CAYLEY.

Note 3. The final claimed efficient presentation for PSL(2,27) on page 49 of [5] is incorrect being, in fact, a presentation for SL(2,27). We thank D.F. Holt for pointing this out.

Note 4. Jamali in [13] gives efficient presentations for most of the maximal subgroups of the simple groups of order less than 10^6.

3. Direct products.

The efficiency of direct products has been of considerable interest for a number of years. First we consider direct squares and, in particular, Wiegold's question [26] as to whether $PSL(2,5) \times PSL(2,5)$ and $SL(2,5) \times SL(2,5)$ are efficient. The first of these questions was answered in the affirmative by Kenne [16] who showed that

$$PSL(2,5) \times PSL(2,5) = \langle x,\ y \mid x^{10} = y^6 = x^4yx^{-1}y^{-3}x^{-1}y^{-1} = (xy^2)^2x^{-1}y^{-1}(xy^{-1})^2 = 1\rangle.$$

The second question was answered in [9] where the balanced efficient presentation

$$SL(2,5) \times SL(2,5) = \langle a,\ b \mid a^2ba^2b^6 = 1,\ a^3b^3 = b^2ab^{-1}a\rangle$$

is given.

We have investigated the direct square $PSL(2,p) \times PSL(2,p)$, p a prime > 3, in an attempt to prove these groups are efficient. We have proved:

Theorem. *For a prime $p > 3$, $p \equiv 1$ (mod 6), $PSL(2,p) \times PSL(2,p)$ has the presentation*

$$\langle a,\ b \mid a^{3p} = b^{3p} = (ab^{-1})^2 = 1,\ (ba^{(p-1)/2}b^{-1}a^{-4})^2 = a^p,$$
$$(ab^{(p-1)/2}a^{-1}b^{-4})^2 = b^p,\ a^{p+1} = b^{p-1}ab^{p-1}\rangle.$$

If $p \equiv -1$ (mod 6) then replace p by $-p$ in the above presentation.

Proof. Let $G = \text{PSL}(2,p) \times \text{PSL}(2,p)$. Then, using a presentation for $\text{PSL}(2,p)$ given in [23],

$$G = \langle x,y,z,t \mid x^2 = y^p = (xy)^3 = (xy^4xy^{(p+1)/2})^2 = 1,$$
$$z^2 = t^p = (zt)^3 = (zt^4zt^{(p+1)/2})^2 = 1,$$
$$[x,z] = [x,t] = [y,z] = [y,t] = 1 \rangle.$$

Put $\alpha = yzt$, $\beta = xyt$. We consider only the case where $p \equiv 1 \pmod 6$. Then $\alpha^3 = y^3$ so $\alpha^{p-1} - y^{p-1} = y^{-1}$ giving $y = \alpha^{1-p}$. Similary $\beta^3 = t^3$ gives $t = \beta^{1-p}$. Since $\alpha\beta^{-1} = xz$ we have $x = \alpha^{1-p}\beta^{-p} = \beta^p\alpha^{p-1}$. Also $z = \alpha^p\beta^{p-1}$.

The relations of G, written in terms of α and β, after simplifying and eliminating obviously redundant relations are:

(i) $(\beta^p\alpha^{p-1})^2 = 1$

(ii) $(\alpha^p\beta^{p-1})^2 = 1$

(iii) $(\beta^p\alpha^3\beta^p\alpha^{(p-1)/2})^2 = 1$

(iv) $(\alpha^p\beta^3\alpha^p\beta^{(p-1)/2})^2 = 1$

(v) $[\alpha^p,\beta^p] = 1$

(vi) $[\alpha^3,\beta^3] = 1$

(vii) $\alpha^{3p} = 1$

(viii) $\beta^{3p} = 1$

First consider (i). Using (v) and (vii) this becomes $\beta^p\alpha^{-1}\beta^p\alpha^{-p-1} = 1$. Hence $\alpha^{p+1} = \beta^p\alpha^{-1}\beta^p$ replaces (i). Similarly $\beta^{p+1} = \alpha^p\beta^{-1}\alpha^p$ replaces (ii). But consider again $\beta^p\alpha^{p-1}\beta^p\alpha^{p-1} = 1$ and this time used (vi) in the form $[\alpha^{p-1},\beta^{p-1}] = 1$. We have

$$\beta^p\alpha^{p-1}\beta^{p-1}\alpha^{p-1} = 1 \Rightarrow \beta^p\beta^{p-1}\alpha^{p-1}\beta\alpha^{p-1} = 1$$

and substituting $\beta^{p+1} = \alpha^p\beta^{-1}\alpha^p$ into this relation gives $\alpha^p\beta^{-1}\alpha^p = \alpha^{p-1}\beta\alpha^{p-1}$ so $(\alpha\beta^{-1})^2 = 1$. This now replaces (ii). Hence, using this to replace $\alpha^{p+1} = \beta^p\alpha^{-1}\beta^p$ by $\alpha^{p+1} = \beta^{p-1}\alpha\beta^{p-1}$ we have new relations

(i) $\alpha^{p+1} = \beta^{p-1}\alpha\beta^{p-1}$

(ii) $(\alpha\beta^{-1})^2 = 1$.

The next step is to show that $[\alpha^p,\beta^p] = 1$ and $[\alpha^3,\beta^3] = 1$ are redundant. After this we simplify (iii) and (iv). Notice that we can still use (v) and (vi) which are consequences of (i) and (ii). Write (iii) as

$$\beta^p\alpha^3\beta^p\alpha^{(p-1)/2}\beta^p\alpha^3\beta^p\alpha^{(p-1)/2} = 1,$$
$$\Rightarrow \beta\alpha^3\beta^{-2}\alpha^{(p-1)/2}\beta\alpha^{(p-1)/2}\beta^{p-1} = 1 \qquad \text{since } 3|(p-1)/2$$
$$\Rightarrow \alpha^3\beta\alpha^{(p-1)/2}\beta\alpha^3\beta\alpha^{(p-1)/2}\beta^{p-3} = 1$$
$$\Rightarrow (\alpha^3\beta\alpha^{(p-1)/2}\beta)^2 = \beta^{4-p}.$$

At this stage we have G generated by α and β subject to the following six relations:

(1) $\alpha^{3p} = 1$

(2) $\beta^{3p} = 1$

(3) $(\alpha\beta^{-1})^2 = 1$

(4) $(\alpha\beta\alpha^{(p-1)/2}\beta)^2 = \beta^{4-p}$

(5) $(\beta^3\alpha\beta^{(p-1)/2}\alpha)^2 = \alpha^{4-p}$

(6) $\alpha^{p+1} = \beta^{p-1}\alpha\beta^{p-1}$.

To obtain the presentation in the Theorem, (4) and (5) must be modified. First the following commutators are easily proved:

$$[\alpha^p, \beta\alpha^3\beta^{-1}] = [\alpha^p, \beta^{-1}\alpha^3\beta] = 1,$$

$$(*)$$

$$[\beta^p, \alpha\beta^3\alpha^{-1}] = [\beta^p, \alpha^{-1}\beta^3\alpha] = 1.$$

To obtain the new relation to replace (4) rewrite (iii) in the form

$$(\beta^{-p-1}\alpha^3\beta\alpha^{(p-1)/2})^2 = 1.$$

Now use $\beta^{-p-1} = \alpha^{-p+1}\beta^{-1}\alpha^{-p+1}$ to obtain $(\beta^{-1}\alpha^{-p+4}\beta\alpha^{(-p+1)/2})^2 = 1$ giving $\beta\alpha^{(p-1/2}\beta^{-1}\alpha^{p-4}\beta\alpha^{(p-1)/2}\beta^{-1}\alpha^{p-1} = 1$. But $3|(p-1)/2$ so, using the commutators $(*)$, we have

$$\beta\alpha^{(p-1)/2}\beta^{-1}\alpha^{-4}\beta\alpha^{(p-1)/2}\beta^{-1}\alpha^{-p-4} = 1$$

giving

$$(\beta\alpha^{(p-1)/2}\beta^{-1}\alpha^{-4})^2 = \alpha^p.$$

Relation (5) is modified in the same way.

This presentation is not efficient since $M(\mathrm{PSL}(2,p) \times \mathrm{PSL}(2,p)) = 2 \times 2$. However we conjecture:

Conjecture. for p a prime > 3, $p \equiv 1 \pmod 6$, $\mathrm{PSL}(2,p) \times \mathrm{PSL}(2,p)$ has the efficient presentation

$$\langle a, b \mid (ab^{-1})^2 = a^{-3p}, \ (ab^{(p-1)/2}a^{-1}b^{-4})^{-2} = b^p,$$

$$(ba^{(p-1)/2}b^{-1}a^{-4})^2 = a^p, \ b^{p+1} = a^{p-1}ba^{p-1}\rangle.$$

If $p \equiv -1 \pmod 6$ replace p by $-p$.

We have verified the conjecture for $p = 5$, ,7, 11, 13 and 17. [**Note added in proof:** We have since shown the conjecture to be true and a proof will be given in "Efficient presentations of the groups $\mathrm{PSL}(2,p) \times \mathrm{PSL}(2,p)$, p prime", C.M.Campbell, E.F.Robertson and P.D.Williams to appear in the J.London Math. Soc.]

The problem of the efficiency of

$$G = \mathrm{PSL}(2, p_1) \times \mathrm{PSL}(2, p_2) \times \cdots \times \mathrm{PSL}(2, p_n)$$

where the p_i are distinct primes > 3 is solved. For, starting from the efficient presentation for $\mathrm{SL}(2, \mathbf{Z}_m)$, $m = p_1 p_2 \cdots p_n$, given in [2], one can add n relations, the ith relation effectively making the central element of order 2 in $\mathrm{SL}(2,p_i)$ equal to 1.

Some work on direct products of groups $\mathrm{PSL}(2,p^{n_i})$ for a fixed prime p and different n_i's is discussed in [10] where some efficient presentations for groups of the form $\mathrm{PSL}(2,q_1) \times \mathrm{PSL}(2,q_2)$, q_1, q_2 prime powers, are also given.

References

[1] F.R. Beyl and J. Tappe, *Group extensions, representations and the Schur multiplicator*, Lecture Notes in Mathematics 958, Springer–Verlag, Berlin, 1982.

[2] C.M. Campbell and E.F. Robertson, 'A deficiency zero presentation for SL(2,p)', *Bull. London Math. Soc.* 12(1980), 17–20.

[3] C.M. Campbell and E.F. Robertson, 'The efficiency of simple groups of order $< 10^5$', *Comm. Algebra* 10(1982), 217–225.

[4] C.M. Campbell and E.F. Robertson, 'Some problems in group presentations', *J. Korean Math. Soc.* 19(1983), 59–64.

[5] C.M. Campbell and E.F. Robertson, 'On a class of groups related to $SL(2,2^n)$', in *Computational Group Theory* (ed. M.D. Atkinson, Academic Press, London, 1984), 43–49.

[6] C.M. Campbell and E.F. Robertson, 'Presentations for the simple groups G, $10^5 < |G| < 10^6$', *Comm. Algebra* 12(1984), 2643–2663.

[7] C.M. Campbell and E.F. Robertson, 'A CAYLEY file of finite simple groups G, $10^5 < |G| < 10^6$', *EUROCAL '85* (Lecture Notes in Computer Science 204, Springer–Verlag, Berlin, 1985), 243–244.

[8] C.M. Campbell and E.F. Robertson, 'Computing with finite simple groups and their covering groups', in *Computers in Algebra* (ed. M.C. Tangora, Marcel Dekker, New York, 1988), 17–26.

[9] C.M. Campbell, T. Kawamata, I. Miyamoto, E.F. Robertson and P.D. Williams, 'Deficiency zero presentations for certain perfect groups', *Proc. Royal Soc. Edinburgh* 103A (1986), 63–71.

[10] C.M. Campbell, E.F. Robertson and P.D. Williams, 'On presentations of $PSL(2,p^n)$', submitted for publication.

[11] J.J. Cannon, J. McKay and K–C. Young, 'The non–abelian simple groups G, $|G| < 10^5$ – presentations', *Comm. Algebra* 7(1979) 1397–1406.

[12] J.H. Conway, R.T. Curtis, S.P. Norton, R.A. Parker and R.A. Wilson, *Atlas of Finite groups: Maximal Subgroups and Ordinary Characters for Simple Groups* (Clarendon Press, Oxford, 1985).

[13] A. Jamali, *Computing with simple groups: maximal subgroups and presentations*, Ph.D. thesis, University of St. Andrews, 1988.

[14] A. Jamali and E.F. Robertson, 'Efficient presentations for certain simple groups', *Comm. Algebra*, to appear.

[15] G. Karpilovsky, *The Schur multiplier* (Oxford University Press, Oxford, 1987).

[16] P.E. Kenne, 'Presentations for some direct products of groups', *Bull. Austral. Math. Soc.* 28(1983), 131–133.

[17] P.E. Kenne, 'Efficient presentations for simple groups', *The Cayley Bulletin* 2(1985), 38.

[18] P.E. Kenne, 'Efficient presentations for three simple groups', *Comm. Algebra* 14(1986), 797–800.

[19] M.A. Kervaire, 'Multiplicateurs de Schur et K–theorie', in *Essays on Topology and Related Topics*, Memoires dédiés à Georges de Rham (eds. A. Haefliger and R. Narasimhan, Springer–Verlag, Berlin, 1970), 212–225.

[20] J. McKay and K–C. Young, 'The non–abelian simple groups G, $|G| < 10^6$ –minimal generating pairs', *Math. Comp.* 33(1979), 812–814.

[21] E.F. Robertson, Efficiency of finite simple groups and their covering groups', *Contemp. Math.* 45(1985), 287–294.

[22] I. Schur, 'Untersuchungen über die Darstellung der endlichen Gruppen durch gebrochene lineare Substitutionen', *J. Reine Angew. Math.* 132(1907), 85–137.

[23] J.G. Sunday, ' Presentations of the groups SL(2,m) and PSL(2,m)', *Canad. J. Math.* 24(1972), 1129–1131.

[24] R.G. Swan,'Minimal resolution for finite groups', *Topology* 4(1965), 193–208.

[25] J.W. Wamsley, *The deficiency of finite groups*, Ph.D. thesis, University of Queensland, 1968.

[26] J. Wiegold, 'The Schur multiplier', in *Groups – St Andrews* 1981 (eds. C.M. Campbell and E.F. Robertson, L.M.S. Lecture Notes 71, Cambridge University Press, Cambridge, 1982), 137–154.

[27] P.D. Williams, *Presentations of linear groups*, Ph.D. thesis, University of St Andrews, 1982.

[28] H.J. Zassenhaus, 'A presentation of the groups PSL(2,p) with three defining relations', *Canad. J. Math.* 21(1969), 310–311.

Department of Mathematical Sciences
University of St Andrews, North Haugh
St Andrews, Fife KY16 9SS
Scotland

*   Department of Mathematics
C.S.U. San Bernardino
5500 University Parkway
San Bernardino, CA 92407
U.S.A.

ON n-GROUPOIDS DEFINED ON FIELDS

Jung Rae Cho

Abstract. Let us define n-ary operations f on the real field $\mathbf{R}$ and the Galois fields $GF(p)$ by $f(x_1, x_2, \cdots, x_n) = \alpha_1 x_1 + \alpha_2 x_2 + \cdots + \alpha_n x_n$ with $\alpha_1, \alpha_2, \cdots, \alpha_n$, none of them zero, in the fields. Then we obtain n-groupoids. Let B be the class of all n-groupoids defined on Galois fields in this way. In this paper, we will show that the equational theory of B is precisely that of $(\mathbf{R}, f)$ if and only if α_1, α_2, $\cdots, \alpha_n$, as elements of $\mathbf{R}$, are algebraically independent. If we restrict our attention only to the case when all α_i's are equal to α, then the equational theory of B is exactly that of $(\mathbf{R}, f)$ if and only if α is transcendental as an element of $\mathbf{R}$. Unlike the groupoid cases ([1], [5]) and the case when $\alpha_1 + \alpha_2 + \cdots + \alpha_n = 1$ ([2]), the basis problems for the equational theories of the above n-groupoids or the above classes are not so simple.

1. Introduction.

An *n-groupoid* is a pair (A, f) of a nonempty set A and an n-ary operation f on A. For simplicity, we will use the notation $[x_1 x_2 \cdots x_n]$ for $f(x_1, x_2, \cdots, x_n)$. An n-groupoid is called *idempotent* if A satisfies the identity $[x_1 x_2 \cdots x_n] = x$ and *medial* if it satisfies the identity

$$[[x_{11} x_{12} \cdots x_{1n}][x_{21} x_{22} \cdots x_{2n}] \cdots [x_{n1} x_{n2} \cdots x_{nn}]]$$

$$= [[x_{11} x_{21} \cdots x_{n1}][x_{12} x_{22} \cdots x_{n2}] \cdots [x_{1n} x_{2n} \cdots x_{nn}]]$$

and *symmetric* if satisfies $[x_1 x_2 \cdots x_n] = [x_{\sigma(1)} x_{\sigma(2)} \cdots x_{\sigma(n)}]$ for every permutation σ on $\{1, 2, \cdots, n\}$.

Given a class B of n-groupoids, the *equational theory* of B is the set of all identities satisfied by every member of B. For an equational theory T and words s, t, we use the notation $s \equiv t \pmod{T}$ if $s = t$ is an identity in T.

We say real numbers $p_1, p_2, \cdots, p_n$ are *algebraically independent* if $(p_1, p_2, \cdots, p_n)$ is not a root of any non-trivial integral coefficient multinomial in n variables. Otherwise, they are said to be *algebraically dependent*.

S. Fajtlowicz and J. Mycielski ([4]) showed that the medial law $(xy)(zw) = (xz)(yw)$ and the idempotent law $xx = x$ form a basis for the equational theory of the groupoid $(\mathbf{R}, o)$, defined on the real field $\mathbf{R}$ by $x \circ y = qx + (1 - q)y$, if and only if q is transcendental. The author ([1]) has shown that the medial and the idempotent laws form a basis for the equational theory of the class of groupoids defined on Galois fields by $x \circ y = qx + (1 - q)y$, where q a nonzero element of the fields. J. Ježek and T. Kepka showed that the medial and the commutative laws (the medial and the cancellative laws, resp.) form a basis for the equational theory of the groupoid defined on $\mathbf{R}$ by $x \circ y = qx + qy$ ($x \circ y = qx + py$, resp.) if q is transcendental (p, q are algebraically independent, resp.). The author ([1]) has shown the converses of these results are also true. The author also has shown that if we define groupoids operations as above on Galois fields with $q \neq 0$ ($q, p \neq 0$, resp), the above pairs of identities form bases for the equational theory of these classes of groupoids.

Generalizing the result of S, Fajtlowicz and J. Mycielski to n-groupoids, the author has shown ([2]) that the medial law and the idempotent law form a basis for the equational theory of the n-groupoid $(\mathbf{R}, f)$, defined on $\mathbf{R}$ by

$$f(x_1, x_2, \cdots, x_n) = \alpha_1 x_1 + \cdots + \alpha_{n-1} x_{n-1} + (1 - \alpha_1 - \cdots - \alpha_{n-1}) x_n$$

with $\alpha_i \in \mathbf{R}$ ($i = 1, 2, \cdots, n - 1$), if and only if $\alpha_1, \alpha_2, \cdots, \alpha_{n-1}$ are algebraically independent. It is also shown by the author that the class of n-groupoids defined on Galois fields as above has exactly the same equational theory. In this paper, we will see how much can be said with operations defined on fields by

$$[x_1 x_2 \cdots x_n] = \alpha_1 x_1 + \alpha_2 x_2 + \cdots + \alpha_n x_n.$$

Let $X = \{x_1, x_2, \cdots, x_n\}$ be a countable set of variables and let W be set of all n-groupoid words in X. We sometimes regard W as the free n-groupoid on X. Throughout this paper, all words will be assume to be in W, and identities are defined in terms of these words.

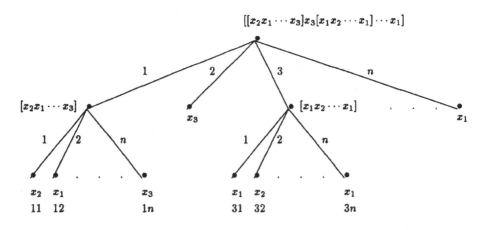

Figure

It is sometimes easy to consider the structure of words by labelled trees ([3], [7]), especially for complicated words likes those of n-groupoids ([2]). An n-ary tree is a tree in which every node except the end nodes has exactly n branches from it. With each n-groupoid word t, we associate a labelled n-ary tree in the natural way as follows: The root is labelled t, and if any node is labelled $[u_1 u_2 \cdots u_n]$ then the nodes branched from it are labelled $u_1, u_2, \cdots, u_n$ from the left. The end nodes are always labelled with variables. We label every i-th branch of each node with i, and hence every node is associated a sequence of $\{1, 2, \cdots, n\}$, which is called the *address* of the node, corresponding to the branches in the path from the root to the node (See Figure on the previous page). In this way, each word can be represented by a labelled tree and its subword by a node of the tree. The *level* of a node (or a subword) is the length of its address. The *depth* of the tree is the maximum level its end nodes can have.

2. Symmetric n-groupoids defined on fields.

For every prime number p and nonzero element α of $\mathrm{GF}(p)$, the Galois field with p elements, we define an n-ary operation $[\ ]_{p\alpha}$ on $\mathrm{GF}(p)$ by

$$[x_1 x_2 \cdots x_n]_{p\alpha} = \alpha x_1 + \alpha x_2 + \cdots + \alpha x_n$$

and let $F_{p\alpha}$ denote the n-groupoid $(\mathrm{GF}(p), [\ ]_{p\alpha})$. It is easily checked that $F_{p\alpha}$ is symmetric and medial. Let

$$B = \{F_{p\alpha} \mid p : \text{prime}, \ \alpha \in \mathrm{GF}(p), \ \alpha \neq 0\}$$

and let T be the equational theory of B. Let $N_i(t; x)$ denote the number of the variable x at level i of the word t, and let $\bar{t}$ denote the term function induced by t on the relevant n-groupoids defined above.

Lemma 2.1. *For every word t, $\bar{t} = \sum_{x \in X} p_x \, x$, where*

$$p_x = \sum_i N_i(t; x) \alpha^i.$$

Proof. By induction on the depth of words.

Theorem 2.2. *Let s and t be words. Then $s \equiv t \pmod{T}$ if and only if $N_i(s; x) = N_i(t; x)$ for every $x \in X$ and i.*

Proof. By the preceding lemma, $\bar{s} = \bar{t}$ on $F_{p\alpha}$ if and only if

$$\sum_i \{N_i(s; x) - N_i(t; x)\} \, \alpha^i \equiv 0 \pmod{p}$$

for all $x \in X$. Thus one direction of the assertion is trivial. Suppose $s \equiv t \pmod{T}$ but $N_i(s; x) \neq N_i(t; x)$ for some i and x. Let m be the bigger between the depths of s and t and choose a prime number $p > n^m$. Then $N_i(s; x) \leq n^m < p$ and $N_i(t; x) \leq n^m < p$, and so $N_i(s; x) \not\equiv N_i(t; x) \pmod{p}$. Then

$$P(z) = \sum_i \{N_i(s; x) - N_i(t; x)\} \, z^i$$

is a non-trivial polynomial in z of degree at most m, and has at most m roots in GF(p). Since $m + 1 \leq n^m < p$, there is a nonzero element α in GF(p) such that $P(\alpha) \neq 0$. That is, $\bar{s} \neq \bar{t}$ on $F_{p\alpha}$, which contradicts the fact $s \equiv t \pmod{T}$.

For any nonzero real number α, define an n-ary operation $[\ ]_\alpha$ on the real field $\mathbf{R}$ by

$$[x_1 x_2 \cdots x_n]_\alpha = \alpha x_1 + \alpha x_2 + \cdots + \alpha x_n,$$

then $(\mathbf{R}, [\ ]_\alpha)$ is a symmetric medial n-groupoid. If $\bar{t}$ denote the term function induced by the word t on $(\mathbf{R}, [\ ]\alpha)$, we can show that

$$\bar{t} = \sum_{x \in X} (\sum_i N_i(t; x) \alpha^i)\, x.$$

Thus, $s = t$ is an identity satisfied by $(\mathbf{R}, [\ ]_\alpha)$ if and only if

$$\sum_i \{N_i(s; x) - N_i(t; x)\}\, \alpha^i = 0$$

for all $x \in X$. If α is transcendental, this is equivalent to having $N_i(s; x) = N_i(t; x)$ for all i and x. Hence we have the following lemma.

Lemma 2.3. *Let α be a transcendental number. Then $s = t$ is an identity satisfied by $(\mathbf{R}, [\ ]_\alpha)$ if and only if $N_i(s; x) = N_i(t; x)$ for all i and x.*

Suppose α is an algebraic number. Then there is an integer m and integers $b_0, b_1, \cdots, b_n$, not all zero, such that

$$b_0 + b_1 \alpha^1 + b_2 \alpha^2 + \cdots + b_m \alpha^m = 0. \tag{1}$$

We choose an integer k such that

$$n^k \geq |b_i| + \frac{1}{n}|b_{i+1}| + \frac{1}{n^2}|b_{i+2}| + \cdots + \frac{1}{n^{m-1}}|b_m|$$

for $i = 0, 1, 2, \cdots, n$. From (1), we have

$$b_0 \alpha^k + b_1 \alpha^{k+1} + b_2 \alpha^{k+2} + \cdots + b_m \alpha^{k+m} = 0. \tag{2}$$

Now we construct two words s and t of depth $m + k$ as follows: if $i < k$ then $N_i(s; x) = N_i(t; x)$ for all x, and at each level $k + i$, $i = 0, 1, \cdots, m$,

(i) s has $|b_i|$ x_1's if $b_i \geq 0$ or $|b_i|$ x_2's if $b_i \leq 0$,

(ii) t has $|b_i|$ x_2's if $b_i \geq 0$ or $|b_i|$ x_1's if $b_i \leq 0$,

(iii) $N_i(s; x) = N_i(t; x)$ for all $x \neq x_1,\ x_2$.

Then, it is easy to check that $N_i(s; x) - N_i(t; x) = 0$ if $i < k$ or $x \neq x_1,\ x_2$, and that $N_{k+i}(s; x_1) - N_{k+i}(t; x_1) = b_i$ and $N_{k+i}(s; x_2) - N_{k+i}(t; x_2) = b_i$ for $i = 1, 2, \cdots, n$. Thus, from (2),

$$\sum_i \{N_{k+i}(s; x) - N_{k+i}(t; x)\}\, \alpha^{k+i} = 0, \quad \text{or} \quad \sum_i \{N_i(s; x) - N_i(t; x)\}\, \alpha^i = 0$$

for all x. Hence, $s = t$ is an identity satisfied by $(R, [\ ]_\alpha)$. However, $N_i(s; x) \neq N_i(t; x)$ if $x = x_1$ or $x = x_2$ for $k < i < m$. That is, $s = t$ is not an identity in T, by Theorem 2.2.

We have shown the following.

Lemma 2.4. *If α is not a transcendental number, then $(R, [\ ]_\alpha)$ satisfies more identities than are in T.*

Example 2.5 Take $\sqrt{2}$ for α. Then the 3-groupoid $(R, [\ ]_{\sqrt{2}})$ satisfies the identity

$$[[x_3 x_3 [x_1 x_3 x_3]] x_2 x_2] = [x_1 [[x_3 x_2 x_3] x_2 x_3] x_1]$$

which is not an identity in T by Theorem 2.2, since $N_1(s; x_1) = 0 \neq 2 = N_1(t; x_1)$.

Combining the preceding two lemmas, we have:

Theorem 2.6. *The equational theory of $(R, [\ ]_\alpha)$ is precisely that of the class B if and only if α is transcendental.*

In groupoid case, that is, the case when $n = 2$, the equational theory of the class B is precisely the set of all consequences of the medial and the symmetric laws ([1]). In that case, the symmetric law is nothing more than the commutative law. Thus, the medial and the commutative laws form a basis for the equational theory of B, and so the equational theory is finitely based. However, this is not the case here for n-groupoids when $n \geq 3$. With a counterexample we now show that there are identities which are in T, but not consequences of the symmetric and the medial laws. Remarkably, the example is a very small n-groupoid, with only three elements.

Example 2.7. Let $A = 1,2,3$ and $[\ ]$ be an n-ary operation, $n \geq 3$, defined on A by

$$[x_1 x_2 \cdots x_n] = \begin{cases} 2 & \text{if only one of } x_i\text{'s is 3 and the rest are 2} \\ 1 & \text{otherwise.} \end{cases}$$

It is easily seen that $(A, [\ ])$ is a symmetric and medial. However, it does not satisfy the identity

$$[[x_1 x_2 \cdots x_n][y_1 y_2 \cdots y_n] z_3 \cdots z_n] = [[y_1 x_2 \cdots x_n][x_1 y_2 \cdots y_n] z_3 \cdots z_n]$$

since

$$[[22 \cdots 2][332 \cdots 2]2 \cdots 23] = [112 \cdots 23] = 1 \neq 2 = [222 \cdots 23] = [[32 \cdots 2][232 \cdots 2]2 \cdots 23].$$

Thus this identity is not a consequence of the symmetric and the medial laws. However, calculating the variables at each level, it is seen to be an identity in T by Theorem 2.2.

Here we have a problem and a question. Give a set (finite or infinite) of identities which is a basis for T. Is T finitely based? Since Theorem 2.2 can serve as a normal form theorem for the variety generated by B, the following is also a related problem: 'Find a concrete normal form theorem for the variety of all symmetric medial n-groupoids'.

3. Cancellative n-groupoids defined on fields.

We call an n-groupoid *cancellative* if it satisfies the quasi-identity

$$[x_1 \cdots x_{i-1} y x_{i+1} \cdots x_n] = [x_1 \cdots x_{i-1} z x_{i+1} \cdots x_n] \implies y = z,$$

for each $i = 1, 2, \cdots, n$.

For every prime number p and nonzero elements α_1, α_2, $\cdots$, α_n of GF(p), we define an n-ary operation $[\cdot]_{p\alpha_1 \cdots \alpha_n}$ on GF(p) by

$$[x_1 x_2 \cdots x_n]_{p\alpha_1 \cdots \alpha_n} = \alpha_1 x_1 + \alpha_2 x_2 + \cdots + \alpha_n x_n$$

and let $F_{p\alpha_1 \cdots \alpha_n}$ denote the n-groupoid $(\text{GF}(p), [\cdot]_{p\alpha_1 \cdots \alpha_n})$. It is easily checked that $F_{p\alpha_1 \cdots \alpha_n}$ is cancellative and medial, but not symmetric in general. Let

$$B = \{F_{p\alpha_1 \cdots \alpha_n} \mid p : \text{prime}, \ \alpha_i \in \text{GF}(p), \ \alpha_i \neq 0 \ (1 \leq i \leq n)\}$$

and let T be the equational theory of B. Let, $N_{j_1 \cdots j_n}(t; x)$ denote the number of the variable x in t at the addresses with j_i i's for $i = 1, 2, \cdots, n$, and let $\bar{t}$ denote the term function induced by t on the relevant n-groupoids defined above.

Lemma 3.1. *For every word t, $\bar{t} = \sum_{x \in X} p_x \, x$, where*

$$p_x = \sum_{j_1, \cdots, j_n} N_{j_1 \cdots j_n}(t; x) \alpha_1^{j_1} \alpha_2^{j_2} \cdots \alpha_n^{j_n}.$$

Proof. By induction on the depth of words.

The following lemma tells about the number of solutions on multinomials on Galois fields.

Lemma 3.2. ([6], p.275) *Let $P(z_1, z_2, \cdots, z_3)$ be a multinomial of degree m in n variables over the Galois field GF(p), then there are at most mp^{n-1} n-tuples in $(GF(p))^n$ which are solutions of $P(z_1, z_2, \cdots, z_n)$.*

Theorem 3.3. *Let s and t be words. Then $s \equiv t$ (mod T) if and only if $N_{j_1 \cdots j_n}(s; x) = N_{j_1 \cdots j_n}(t; x)$ for every $x \in X$ and $j_1, \cdots, j_n$.*

Proof. By the preceding lemma, $\bar{s} = \bar{t}$ on $F_{p\alpha_1 \cdots \alpha_n}$ if and only if

$$\sum_{j_1, \cdots, j_n} \{N_{j_1 \cdots j_n}(s; x) - N_{j_1 \cdots j_n}(t; x)\} \, \alpha_1^{j_1} \alpha_2^{j_2} \cdots \alpha_n^{j_n} \equiv 0 \pmod{p}$$

for all $x \in X$. Hence half of the assertion is trivial. Suppose $s \equiv t$ (mod T) but $N_{j_1 \cdots j_n}(s; x) \neq N_{j_1 \cdots j_n}(t; x)$ for some $j_1, \cdots, j_n$ and x. Let m be the bigger of the depths of s and t and choose a prime number $p > n^m$. Then $N_{j_1 \cdots j_n}(s; x) \leq n^m < p$ and $N_{j_1 \cdots j_n}(t; x) \leq n^m < p$, and so $N_{j_1 \cdots j_n}(s; x) \not\equiv N_{j_1 \cdots j_n}(t; x) \pmod{p}$. Then

$$P(z_1, z_2, \cdots, z_n) = \sum_{j_1, \cdots, j_n} \{N_{j_1 \cdots j_n}(s; x) - N_{j_1 \cdots j_n}(t; x)\} \, z_1^{j_1} z_2^{j_2} \cdots z_n^{j_n}$$

is a non-trivial polynomial in $z_1, z_2, \cdots, z_n$ of degree at most m, and has at most mp^{n-1} n-tuples of solutions, by Lemma 3.2. On the other hand, there are less than np^{n-1} n-tuples in $(GF(p))^n$ with at least one zero for its components. However, $p^n = pp^{n-1} > n^m p^{n-1} \geq (n + m)p^{n-1}$. Thus there are nonzero $\alpha_1, \alpha_2, \cdots \alpha_n$ in $GF(p)$ such that $P(\alpha_1, \alpha_2, \cdots \alpha_n) \neq 0$. That is, $\bar{s} \neq \bar{t}$ on $F_{p\alpha_1 \cdots \alpha_n}$, which contradicts the fact $s \equiv t \pmod{T}$.

For any nonzero real numbers $\alpha_1, \alpha_2, \cdots, \alpha_n$, define an n-ary operation $[\ ]_{\alpha_1 \cdots \alpha_n}$ on the real field $\mathbf{R}$ by

$$[x_1 x_2 \cdots x_n]_{\alpha_1 \cdots \alpha_n} = \alpha_1 x_1 + \alpha_2 x_2 + \cdots + \alpha_n x_n,$$

then $(\mathbf{R}, [\ ]_{\alpha_1 \cdots \alpha_n})$ is a cancellative medial n-groupoid. If $\bar{t}$ denote the term function induced by the word t on $(\mathbf{R}, [\ ]_{\alpha_1 \cdots \alpha_n})$, we can show that

$$\bar{t} = \sum_{x \in X} \left(\sum_{j_1, \cdots, j_n} N_{j_1 \cdots j_n}(t; x) \alpha_1^{j_1} \alpha_2^{j_2} \cdots \alpha_n^{j_n} \right) x.$$

Thus, $s = t$ is an identity satisfied by $(\mathbf{R}, [\ ]_{\alpha})$ if and only if

$$\sum_i \{N_{j_1 \cdots j_n}(s; x) - N_{j_1 \cdots j_n}(t; x)\} \, \alpha_1^{j_1} \alpha_2^{j_2} \cdots \alpha_n^{j_n} = 0$$

for all $x \in X$. If $\alpha_1, \alpha_2, \cdots, \alpha_n$ are algebraically independent, this is equivalent to having $N_{j_1 \cdots j_n}(s; x) = N_{j_1 \cdots j_n}(t; x)$ for all $j_1, j_2, \cdots, j_n$ and x. Hence we have the following lemma by Theorem 3.3.

Lemma 3.4. *Let $\alpha_1, \alpha_2, \cdots, \alpha_n$ be algebraically independent Then $s = t$ is an identity satisfied by $(\mathbf{R}, [\ ]_{\alpha_1 \cdots \alpha_n})$ if and only if $N_{j_1 \cdots j_n}(s; x) = N_{j_1 \cdots j_n}(t; x)$ for all $j_1, \cdots, j_n$ and x.*

Suppose $\alpha_1, \alpha_2, \cdots, \alpha_n$ are algebraically dependent. Then there are integers $b_{j_1 \cdots j_n}$'s, not all zero, such that

$$\sum_{j_1, \cdots, j_n} b_{j_1 \cdots j_n} \alpha_1^{j_1} \alpha_2^{j_2} \cdots \alpha_n^{j_n} = 0. \tag{3}$$

We choose an integer $q_1, q_2, \cdots, q_n$ such that

$$\binom{q_1 + j_1 + \cdots + q_n + j_n}{j_1, \cdots, j_n} \geq \sum_{j_1, \cdots, j_n} |b_{j_1 \cdots j_n}|$$

for every $j_1, \cdots, j_n$, where

$$\binom{q_1 + j_1 + \cdots + q_n + j_n}{j_1, \cdots \cdots, j_n}$$

is a multinomial coefficient. Let $k_i = \max\{j_i \mid b_{j_1 \cdots j_i \cdots j_n} \neq 0\}$ for each $i = 0, 1, 2, \cdots, n$. For example, for $\alpha_1^3 \alpha_2 + \alpha_1 \alpha_2^2 = 0$, $k_1 = 3$ and $k_2 = 2$. Now let s and t be words of depth $q_1 + k_1 + \cdots q_n + k_n$

constructed as follows: if $j_i < q_i$ for some i then $N_{j_1 \cdots j_n}(s; x) = N_{j_1 \cdots j_n}(t; x)$ for all x, and at addresses with $q_1 + j_i$ i's for $i = 1, 2, \cdots, n$,

(i) s has $|b_{j_1 \cdots j_n}|$ x_1's if $b_{j_1 \cdots j_n} > 0$, or x_2's otherwise.

(ii) t has $|b_{j_1 \cdots j_n}|$ x_2's if $b_{j_1 \cdots j_n} > 0$, or x_1's otherwise.

(iii) $N_{q_1 + j_1 \cdots q_n + j_n}(s; x) = N_{q_1 + j_1 \cdots q_n + j_n}(t; x)$ if $x \neq x_1, x_2$.

Then, for every $j_1, j_2, \cdots, j_n$,

$$N_{q_1 + j_1 \cdots q_n + j_n}(s; x) - N_{q_1 + j_1 \cdots q_n + j_n}(t; x) = \begin{cases} b_{j_1 \cdots j_n} & \text{if } x = x_1 \\ -b_{j_1 \cdots j_n} & \text{if } x = x_2 \\ 0 & \text{otherwise.} \end{cases}$$

Since $N_{j_1 \cdots j_n}(s; x) - N_{j_1 \cdots j_n}(t; x) = 0$ if $j_i < q_i$ for some i, we have, by (3),

$$\sum_{j_1, \cdots, j_n} \{ N_{j_1 \cdots j_n}(s; x) - N_{j_1 \cdots j_n}(t; x) \} \alpha_1^{j_1} \cdots \alpha_n^{j_n}$$

$$= \{ \sum_{j_1, \cdots, j_n} \{ N_{q_1 + j_1 \cdots q_n + j_n}(s; x) - N_{q_1 + j_1 \cdots q_n + j_n}(t; x) \} \alpha_1^{j_1} \cdots \alpha_n^{j_n} \} \, \alpha_1^{q_1} \cdots \alpha_n^{q_n}$$

$$= 0$$

for all $x \in X$. Hence $s = t$ is an identity satisfied by $(\mathbf{R}, [\ ]_{\alpha_1 \cdots \alpha_2})$. However, it is not an identity in T by Theorem 3.3. Hence we have:

Lemma 3.5. *If $\alpha_1, \alpha_2, \cdots, \alpha_n$ are not algebraically independent, then $(\mathbf{R}, [\ ]_{\alpha_1 \cdots \alpha_n})$ satisfies more identities than are in T.*

Example 3.6. If we take π, e, and $\sqrt{\pi + 2e}$ for α_1, α_2 and α_3, respectively. The 3-groupoid defined by

$$[x, y, z] = \pi\, x + e\, y + \sqrt{\pi + 2e}\, z$$

satisfies the identity

$$[[x_1 x_1 [x_3 x_4 x_2]][x_1 x_3 x_3][x_3 x_4 x_5]] = [[x_2 x_2 [x_3 x_4 x_1]][x_2 x_3 x_4][x_3 x_3 x_5]],$$

which is easily checked not to be an identity in T by Theorem 3.3.

Joining the preceding two lemmas, we have:

Theorem 3.7. *The equational theory of $(\mathbf{R}, [\ ]_{\alpha_1 \cdots \alpha_n})$ is precisely that of the class B if and only if $\alpha_1, \alpha_2, \cdots, \alpha_n$ are algebraically independent.*

For groupoids, that is, n-groupoids where $n = 2$, any identity of the class of groupoids defined on Galois field can be derived from the medial law and the cancellation property ([1],[5]). However, its equational theory is not finitely based ([5]), that is, the class of all cancellative medial groupoid satisfies infinitely many independent identities. It is not known yet that the equational theory for

n-groupoid ($n \geq 3$) case is finitely based or not. It is also yet to be settled whether or not, when $n \geq 3$, the equational theory of the class of n-groupoids defined in this section is the same as that of the class of all cancellative medial n-groupoids.

For the variety of n-groupoids generated by the n-groupoids defined in this section, Theorem 3.3 can serve a normal form theorem. Thus, a related question to above problem is: 'Can we find an easy normal form theorem for the variety generated by all cancellative medial n-groupoid'.

References

[1] J. Cho, 'Varieties of medial groupoids generated by groupoids defined on fields', *Houston J. Math.*, to appear.

[2] J. Cho, 'Idempotent medial n-groupoids defined on fields', *Alg. Univ.*, 25(1988), 235-246.

[3] T. Evans, 'A decision problem for transformation of trees', *Canad. J. Math.*, 15(1963), 584-590.

[4] F. Fajtlowicz and J. Mycielski, 'On convex linear forms', *Alg. Univ.*, 4(1974), 273-281.

[5] J. Ježek and T. Kepka, *Medial Groupoids*, A monograph of Academia Praha, 1983.

[6] R. Lidl and H. Nietherreiter, *Finite fields*, Encyclopedia of Math. and its Appl., Addison-Wesley, 1983.

[7] H. Minc, 'Polynomials and bifurcating root-trees', *Proc. Roy. Soc. Edinburgh*, Sec. A, 64(1957), 319-341.

Department of Mathematics
Pusan National University
Pusan, 609-735
Republic of Korea

AUTOMORPHISMS OF CERTAIN RELATIVELY FREE SOLVABLE GROUPS

C. Kanta Gupta

Let $F_{n,c} = \langle x_1, x_2, \cdots, x_n \rangle$, $n \geq 2$, $c \geq 2$ denote the free nilpotent group of class c freely generated by the set $\{x_1, x_2, \cdots, x_n\}$. Then $F_{n,c} \cong F_n/\gamma_{c+1}(F_n)$ where $F_n = \langle f_1, f_2, \cdots, f_n; \phi \rangle$ is the absolutely free group on $\{f_1, f_2, \cdots, f_n\}$. We denote by T the tame automorphisms of $F_{n,c}$, i.e. the subgroup of $\mathrm{Aut}(F_{n,c})$ induced by $\mathrm{Aut}(F_n)$. It is well-known (see, for instance, Magnus, Karrass and Solitar [8], chapter 3) that $T = \langle \tau_1, \ \tau_2, \ \tau_3, \tau_4 \rangle$, where τ_i are defined by

$$\tau_1 : x_1 \longrightarrow x_1^{-1}, \ x_i \longrightarrow x_i, \ i \neq 1; \quad \tau_2 : x_1 \longrightarrow x_1 x_2, \ x_i \longrightarrow x_i, \ i \neq 1;$$

$$\tau_3 : x_1 \longrightarrow x_2, \ x_2 \longrightarrow x_1, \ x_i \longrightarrow x_i, \ i \neq 1, 2; \quad \tau_4 : x_1 \longrightarrow x_2, \ \cdots, \ x_n \longrightarrow x_1.$$

Since every automorphism of the free abelian group $F_{n,c}/\gamma_2(F_{n,c})$ is tame (see Magnus et al), it follows that $\mathrm{Aut}(F_{n,c}) = \langle T, \mathrm{IA-Aut}(F_{n,c}) \rangle$, where $\mathrm{IA-Aut}(F_{n,c})$ consists of the so-called IA-automorphisms of the form $x_i \longrightarrow x_i d_i$, $i = 1, \cdots, n$, with $d_i \in \gamma_2(F_{n,c})$, the derived subgroup of $F_{n,c}$. In fact, since every automorphism of $F_n/\gamma_3(F_n)$ is tame, the above set can be further modified to automorphisms of the form $x_i \longrightarrow x_i d_i$, $d_i \in \gamma_3(F_{n,c})$. Andreadakis [1] has shown that, for $c \geq 3$, there are IA-automorphisms of $F_{n,c}$, which are not tame. Using the tame automorphisms, it is routine to verify that the generating set for $\mathrm{Aut}(F_{n,c})$ can be reduced to:

$$\mathrm{Aut}(F_{n,c}) = \langle T, \{x_1 \longrightarrow x_1 [x_1, x_{i2}, \cdots, x_{im}], \ x_i \longrightarrow x_i, \ i \neq 1\}, \ 3 \leq m \leq c \rangle, \tag{1}$$

where $x_{ij} \in \{x_1, x_2, \cdots, x_n\}$.

It is natural to raise the following question:

Problem. Together with T how many IA-automorphisms of $F_{n,c}$ are required to generate $\mathrm{Aut}(F_{n,c})$?

[If F is of countable infinite rank then Bryant and Macedonska [5] have proved that $\mathrm{Aut}(F_{\infty,c}) = \langle T \rangle$.]

Goryaga [6] proved that if $n \geq 3 \cdot 2^{c-2} + c$, then $\mathrm{Aut}(F_{n,c}) = \langle T, \theta_3, \cdots, \theta_c \rangle$, where θ_k are defined by

$$\theta_k = \begin{cases} x_1 \longrightarrow x_1 [x_1, x_2, \cdots, x_k] \\ x_i \longrightarrow x_i, \quad i \neq 1. \end{cases} \tag{2}$$

Andreadakis [2] reduced the restriction on n in Goryaga's result significantly by proving that the same conclusion holds for $n \geq c$, i.e. for $n \geq c$ $\mathrm{Aut}(F_{n,c}) = \langle T, \theta_3, \cdots, \theta_c \rangle$.

In joint work with Bryant we further improve Andreadakis' result by eliminating all but one θ_k. We prove

Theorem. (C.K. Gupta and Bryant). *If $n \geq c$, then $\mathrm{Aut}(F_{n,c}) = \langle T, \theta_3 \rangle$.*

Outline of proof. We set $K = \langle T, \theta_3 \rangle$. By Andreadakis [2], we already know that

$$\mathrm{Aut}(F_{n,c}) = \langle T, \theta_3, \cdots, \theta_c \rangle. \tag{3}$$

Thus it suffices to prove that each θ_k lies in K. Choose

$$\alpha = \alpha_{34} = \{x_3 \longrightarrow x_3 x_4, \ x_i \longrightarrow x_i, \ i \neq 3\} \in T \leq K,$$
$$\beta = \beta_{34} = \{x_3 \longrightarrow x_4, \ x_4 \longrightarrow x_3, \ x_i \longrightarrow x_i, \ i \neq 3,4\} \in T \leq K.$$

Then

$$\theta_4^* = (\alpha^{-1}\theta_3\alpha)(\beta^{-1}\theta_3\beta)^{-1}(\theta_3)^{-1} \in K,$$

and θ_4^* has the form

$$\theta_4^* = \begin{cases} x_1 \longrightarrow x_1[x_1, x_2, x_3, x_4]d_5 \\ x_i \longrightarrow x_i, \quad i \neq 1, \end{cases}$$

where $d_5 = d_5(x_1, x_2, x_3, x_4)$ is a product of commutators of weight at least 5 in which x_2 appears at least twice and each other variable appears at least once.

Next, choose

$$\alpha = \alpha_{45} = \{x_4 \longrightarrow x_4 x_5, \ x_i \longrightarrow x_i, \ i \neq 4\} \in T \leq K,$$
$$\beta = \beta_{45} = \{x_4 \longrightarrow x_5, \ x_5 \longrightarrow x_4, \ x_i \longrightarrow x_i, \ i \neq 4,5\} \in T \leq K.$$

Then, as before,

$$\theta_5^* = (\alpha^{-1}\theta_4^*\alpha)(\beta^{-1}\theta_4^*\beta)^{-1}(\theta_4^*)^{-1} \in K,$$

and θ_5^* has the form

$$\theta_5^* = \begin{cases} x_1 \longrightarrow x_1[x_1, x_2, x_3, x_4, x_5]d_6 \\ x_i \longrightarrow x_i, \quad i \neq 1, \end{cases}$$

where $d_6 = d_6(x_1, x_2, x_3, x_4, x_5)$ is a product of commutators of weight at least 6 in which x_2 appears at least twice and each of the other variables appear at least once. Repeating the argument we finally arrive at the set of automorphisms $\theta_4^*, \theta_5^*, \cdots, \theta_c^* \in K$. Since $F_{n,c}$ is nilpotent of class c, we have $d_{c+1} = 1$; and it follows that $\theta_c^* = \theta_c \in K$. Let μ_k, $k = 3, \cdots, c$, denote an arbitrary automorphism of the form

$$\mu_k = \begin{cases} x_1 \longrightarrow x_1 d_k, \quad d_k \in \gamma_k(F_{n,c}) \\ x_i \longrightarrow x_i, \quad i \neq 1. \end{cases} \tag{4}$$

Then, together with T, each μ_c is a consequence of θ_c, and hence $\mu_c \in K$. Now

$$\theta_{c-1}^* = \theta_{c-1}\mu_c,$$

and, since θ_{c-1}^*, $\mu_c \in K$, it follows that $\theta_{c-1} \in K$; and, in turn, as before $\mu_{c-1} \in K$. Thus by reverse induction it follows, using $\theta_k^* = \theta_k\mu_{k+1}$, that each $\theta_k \in K$, $4 \leq k \leq c$, as was to be proved.

Some further improvements to above theorem are possible. We refer to Bryant and Gupta [4] for details.

Theorem. (Bryant and C.K. Gupta [4]). *For $n \geq c - 1 \geq 2$, $\mathrm{Aut}(F_{n,c}) = \langle T, \delta_3 \rangle$, where*

$$\delta_3 = \{x_1 \longrightarrow x_1[x_1, x_2, x_1], \quad x_i \longrightarrow x_i, i \neq 1\}.$$

For $n \leq c - 2$, more IA-automorphisms seem to be required to generate $\mathrm{Aut}(F_{n,c})$. For instance, if $n \geq c - 2 \geq 2$ then $\mathrm{Aut}(F_{n,c}) = \langle T, \delta_3, \delta_4 \rangle$, where

$$\delta_4 = \{x_1 \longrightarrow x_1[x_1, x_2, x_1, x_1], \quad x_i \longrightarrow x_i, i \neq 1\}.$$

For $(c+1)/2 \leq n \leq c-1$, we give a specific generating set which depends only on the difference $c - n$ (see [4]). However, for $n \leq c/2$ we do not know any reasonably small generating set for $\mathrm{Aut}(F_{n,c})$.

Automorphisms of free nilpotent and metabelian groups. Let $M_{n,c}$ denote the free metabelian nilpotent of class c group freely generated by $\{x_1, x_2, \cdots, x_n\}$. Then $M_{n,c} \cong F_n/\gamma_{c+1}(F_n)F_n''$, where, as before, F_n is the absolutely free group on the set $\{f_1, f_2, \cdots, f_n\}$. For $c \geq 3$, a complete description of IA-$\mathrm{Aut}(M_{2,c})$ in terms of generators and defining relations has been given by C.K. Gupta [7]. Once again we seek a reasonable generating set for $\mathrm{Aut}(M_{n,c})$. Jointly with Andreadakis, we have the following most recent results.

Theorem. (Andreadakis and C.K. Gupta [3])
(i) *If $[c/2] \leq n \leq c$, then $\mathrm{Aut}(M_{n,c}) = \langle T, \delta_3, \cdots, \delta_c \rangle$, where*

$$\delta_k = \begin{cases} x_1 \longrightarrow x_1[x_1, x_2, x_1, \cdots, x_1] & (k - 2 \text{ repeats of } x_1) \\ x_i \longrightarrow x_i, & i \neq 1. \end{cases}$$

(ii) *If $n \geq 2$ and $c \geq 3$ then for each $\alpha \in \mathrm{Aut}(M_{n,c})$ there exists a positive integer $a(\alpha)$ such that $\alpha^{a(\alpha)} \in \langle T, \delta_3, \cdots, \delta_c \rangle$ where, in addition, the prime factorization of $a(\alpha)$ uses primes dividing $[c/2]!$.*

Proof. Full details of the proof will be published elsewhere. Here we outline the proof for the case $n = 2$, $c = 5$, which is the first case not covered in the preceding sections.

For $p, q \geq 1$, $p + q - 5$, define

$$\begin{aligned} \delta(p, q) &= x_1 \rightarrow x_1[x_1, x_2, (p-1)x_1, (q-1)x_2] \\ x_2 &\rightarrow x_2. \end{aligned}$$

Thus, for instance, $\delta_5 = \delta(4, 1)$. By induction we only need to prove that some suitable power of each $\delta(p, q)$ lies in the gorup $K = \langle T, \delta_3, \delta_4, \delta_5 \rangle$. Denote by τ_{12} the tame automorphism defined by

$$x_1 \quad \rightarrow \quad x_1 x_2$$
$$x_2 \quad \rightarrow \quad x_2.$$

Since $\delta(4,1)$, $\tau_{12} \in K$, it follows that

$$[\delta(4,1), \tau_{12}], \ [\delta(4,1), \tau_{12}, \tau_{12}], \ [\delta(4,1), \tau_{12}, \tau_{12}, \tau_{12}] \in K.$$

These yield, in turn, the relations

$$\delta(3,2)^3 \delta(2,3)^3 \delta(1,4) \in K, \quad \delta(2,3)^6 \delta(1,4)^6 \in K, \quad \delta(1,4)^6 \in K.$$

which yield the relations

$$\delta(1,4)^6 \in K, \quad \delta(2,3)^6 \in K, \quad \delta(3,2)^{18} \in K,$$

as was to be proved.

References

[1]   S. Andreadakis, 'On the automorphisms of free groups and free nilpotent groups', *Proc. London Math. Soc.* 15(1965), 239–269.

[2]   S. Andreadakis, 'Generators for Aut G, G free nilpotent', *Arch. Math.* 42(1984), 296–300.

[3]   S. Andreadakis and C.K. Gupta, 'Automorphisms of free metabelian nilpotent groups', (under preparation).

[4]   R.M. Bryant and C.K. Gupta, 'Automorphism groups of free nilpotent groups', *Arch. Math.* (to appear).

[5]   Roger M. Bryant and Olga Macedonska, 'Automorphisms of relatively free nilpotent groups of infinite rank', *J. Algebra* (to appear).

[6]   A.V. Goryaga, 'Generators of the automorphism group of a free nilpotent group', *Algebra and Logic* 15(1976), 289–292. [English Translation]

[7]   C.K. Gupta, 'IA-automorphisms of two generator metabelian groups', *Arch. Math.* 37(1981), 106–112.

[8]   W. Magnus, A. Karrass and D. Solitar, *Combinatorial group theory*, Interscience Publ., New York, 1966.

Department of Mathematics,
University of Manitoba,
Winnipeg R3T 2N2,
Canada

HIGHER DIMENSION SUBGROUPS

Narain Gupta

1. Notation, Definition and History.

ZG : integral group ring of a finitely generated group G.

$\Delta(G)$: the augmentation ideal of ZG $[\Delta(G) = ZG(G-1) = \sum_i n_i(g_i - 1),\ n_i \in Z,\ g_i \in G]$.

$D_n(G)$: the n-th dimension subgroup of G $[= G \cap (1 + \Delta^n(G)) = \{g \in G; g - 1 \in \Delta^n(G)\}]$.

$\gamma_n(G)$: the n-th term of the lower central series of G $[$ = the normal closure of all left-normed commutators $[g_1, g_2, \cdots, g_n],\ g_i \in G]$.

Since $[g, h] - 1 = g^{-1}h^{-1}gh - 1 = g^{-1}h^{-1}(gh - hg) = g^{-1}h^{-1}((g-1)(h-1) - (h-1)(g-1)) \in \Delta^2(G)$, a simple induction on $n \geq 2$ shows that $[g_1, g_2, \cdots, g_n] - 1 \in \Delta^n(G)$, implying $\gamma_n(G) \leq D_n(G)$.

A Fundamental Theorem (Magnus, 1937). *If G is a free group then $D_n(G) = \gamma_n(G)$ for all* $n \geq 1$.

[Also attributed to Grün (1936) and Witt(1937). Frank Röhl (1985) has given an account of Grün's work pointing out some errors in his work. I refer to my monograph "Free Group Rings" [Gupta (1987)] for details of the fundamental theorem and other references.]

Dimension Subgroup Conjecture: $D_n(G) = \gamma_n(G),\ \forall n,\ \forall G$.

[The conjecture holds for $n \leq 3$. For $n = 1, 2$, it is trivial. For $n = 3$ it is a well-known result attributed to Higman and Rees (unpublished). It has been re-proved by several authors: Passi (1968), Hoare (1969), Sandling (1972), Losey (1974), Bachmann and Gruenenfelder (1972), Gupta (1982). It is not difficult to see that if the conjecture is false then it is already false for some finite p-group (see, for instance, Passi (1979)). If G is a finite p-group, p odd, then Passi (1968) proved that $D_4(G) = \gamma_4(G)$. Rips (1972) constructed a 2-group counter-example to the conjecture for $n = 4$.]

Counter-example (Rips (1972)): There exists a finite 2-group G such that $D_4(G) \neq \gamma_4(G)$.

Dimension Subgroup Problem: Find the structure of the quotient groups $D_n(G)/\gamma_n(G)$.

[Tahara (1977), (1981) has obtained a solution of the probelm for $n = 4$ and some partial results for $n = 5$.]

The best-known general result towards the problem is :

Theorem (Sjogren, 1979): *For $n \geq 3$, the exponent of $D_n(G)/\gamma_n(G)$ divides a certain number c_n whose primary decomposition uses primes not exceeding $n - 2$.*

[Earlier, Losey (1974) had proved this result for $n = 4$.]

Among the various consequences of Sjogren's theorem are:

Corollary. *If G is a p-group then $D_n(G) = \gamma_n(G)$, $\forall n \leq p + 1$.*

[Earlier, Moran (1970) had proved this result for $n \leq p$.]

Corollary (Hall - Jennings). *If all lower central factors of G are torsion free then $D_n(G) = \gamma_n(G)$.*

Since Rips' counter-example is a metabelian group, it is significant to restrict the study of the dimension subgroup problem to *finitely generated metabelian groups*. Some of what is known is:

Theorem (Gupta, Hales and Passi 1984). *If G is a finitely generated metabelian group then there exists an integer $n_0 = n_0(G/G')$ such that $D_n(G) = \gamma_n(G)$ for all $n \geq n_0$.*

Theorem (Gupta and Tahara 1985). *If G is metabelian p-group, p odd, then $D_n(G) = \gamma_n(G)$ for all $n \leq p + 2$.*

[Note: This is an improvement over Sjogren's result.]

Theorem (Gupta 1988). *If G is a metabelian p-group, then $D_n(G) = \gamma_n(G)$ for all $n \leq 2p - 1$.*

The General Problem —— a reformulation.

Let G be a finite group given by a pre-abelian free presentation

$$1 \longrightarrow R \longrightarrow F \longrightarrow G \longrightarrow 1$$

such that

$$F = \langle x_1, \cdots, x_m; \phi \rangle, \quad m \geq 2,$$

is free and R is the normal closure

$$R = \langle r_1, \cdots, r_m, r_{m+1}, \cdots, r_q \rangle^F,$$

where $e(m)|e(m-1)| \cdots |e(1) > 0$, $r_i = x_i^{e(i)} \zeta_i$, $\zeta_i \in F'$, for $1 \leq i \leq m$, and $r_j \in F'$, for $m+1 \leq j \leq q$. [Such a presentation is always possible. See Magnus, Karrass and Solitar (1966)] Denote by $\mathbf{Z}F$, the integral group ring of F and define ideals $\mathfrak{f}$, $\mathfrak{r}$ of $\mathbf{Z}F$ as follows:

$$\begin{aligned} \mathfrak{f} &= \mathbf{Z}F(F - 1), \quad \text{the augmentation ideal of } \mathbf{Z}F; \\ \mathfrak{r} &= \mathbf{Z}F(R - 1), \quad \text{the ideal of } \mathbf{Z}F \text{ relative to } R. \end{aligned}$$

Then, for each $n \geq 1$,

$$F \cap (1 + \mathfrak{r} + \mathfrak{f}^n) = \{w \in F; w - 1 \in \mathfrak{r} + \mathfrak{f}^n\}$$

is a normal subgroup of F and, in the language of free group rings, the dimension subgroup probelem is equivalent to the identification of $F \cap (1 + \mathfrak{r} + \mathfrak{f}^n)$. We wish to study the quotient groups $F \cap (1 + \mathfrak{r} + \mathfrak{f}^n)/R\gamma_n(F)$.

Clearly, $R\gamma_n(F) \leq F \cap (1 + \mathfrak{r} + \mathfrak{f}^n)$. If w is an arbitrary element of $F \cap (1 + \mathfrak{r} + \mathfrak{f}^n)$, then $w - 1 \in \mathfrak{r} + \mathfrak{f}^n$ implies that modulo $\mathfrak{f}\mathfrak{r} + \mathfrak{f}^n$,

$$w - 1 \equiv \sum_{1 \leq i \leq q} n(i)(r_i - 1) \equiv \left(\prod_{1 \leq i \leq q} r_i^{n(i)} - 1 \right) = r - 1, \quad \text{for some } r \text{ in } R.$$

It follows that $wr^{-1} - 1 \in \mathfrak{f}\mathfrak{r} + \mathfrak{f}^n$. Thus we have the following reduction lemma.

Lemma 1. $F \cap (1 + \mathfrak{r} + \mathfrak{f}^n) = R\gamma_n(F)$ if and only if $F \cap (1 + \mathfrak{f}\mathfrak{r} + \mathfrak{f}^n) \leq R\gamma_n(F)$.

Next we introduce in $\mathbf{Z}F$ a family of ideals as follows:

$$\mathfrak{r}\,(k) = \sum_{i+j+1=k} \mathfrak{f}^i \mathfrak{r} \mathfrak{f}^j, \quad k \geq 2.$$

[Thus $\mathfrak{r}(2) = \mathfrak{f}\mathfrak{r} + \mathfrak{r}\mathfrak{f}$.]

We also introduce a family of normal subgroups of F as follows:

$$R(k) = [R, {}_{k-1} F] = [R, F, \cdots, F], \ k \geq 2, \quad (F \text{ repeats } k - 1 \text{ times})$$

Using the three subgroup lemma $[A, B, C] \leq [B, C, A][C, A, B]$ for normal subgroups A, B, C of F, it is easily seen that

$$[F, \cdots, F, R, {}_j F] \leq R(k), \text{ for } i + j + 1 = k, \ i, j \geq 0, \quad (F \text{ repeats } i \text{ times}).$$

For each $k \geq 2$, denote

$$D_n(k, R) = F \cap (1 + \mathfrak{r}(k) + \mathfrak{f}^n).$$

[Note: $R(k)\gamma_n(F) \leq D_n(k, R) \leq \{D_{n-1}(k, R), D_n(k - 1, R)\}$]

Dimension Subgroup Problem   (revised version): Identify the quotients
$$D_n(k, R)/R(k)\gamma_n(F) \text{ for each } 2 \leq k \leq n.$$

[Note that the case $k = 2$ is the familiar version of the problem. Sjogren (1979) proved that $D_n(n - 1, R) = R(n - 1)\gamma_n(F)$.]
We refer to subgroups $D_n(k, R)$ as *the higher dimension subgroups* of F relative to R.

2. Dimension Subgroups of Metabelian groups.

Let G be a finite metabelian group given by a pre–abelian free presentation

$$1 \longrightarrow R \longrightarrow F \longrightarrow G \longrightarrow 1$$

such that $F = \langle x_1, \cdots, x_m; \phi \rangle$, $m \geq 2$, is free and R is the normal closure

$$R = \langle r_1, \cdots, r_m, r_{m+1}, \cdots, r_q \rangle^F F'',$$

where $e(m)|e(m-1)| \cdots |e(1) > 0$, $r_i = x_i^{e(i)}\zeta_i$, $\zeta_i \in F'$, for $1 \leq i \leq m$, and $r_j \in F'$, for $m+1 \leq j \leq q$.

Let $S = RF' = \langle x_1^{e(1)}, \cdots, x_m^{e(m)} \rangle F'$. Then F/S is abelian and we have $S' \leq R \leq S \leq F$. Thus from Lemma 1 we deduce the following theorem for the dimension subgroups of metabelian groups.

Theorem. *If* $F \cap (1 + \mathfrak{f}\mathfrak{s} + \mathfrak{f}^n) = S'\gamma_n(F)$ *then* $F \cap (1 + \mathfrak{r} + \mathfrak{f}^n) = R\gamma_n(F)$, *where* $\mathfrak{s} = \mathbf{Z}F(S-1)$.

[The theorems of Gupta, Hales and Passi (1984), Gupta and Tahara (1985) have been obtained using this reduction theorem.]

The counter-example of Rips further shows that, in general, $F \cap 1 + \mathfrak{f}\mathfrak{s} + \mathfrak{f}^n) \neq S'\gamma_n(F)$ for $n \geq 4$. In view of the facts that

$$F \cap (1 + \mathfrak{f}\mathfrak{s}) = S' \quad \text{(Magnus 1939)}$$

and

$$F \cap (1 + \mathfrak{f}^n) = \gamma_n(F) \quad \text{(Magnus 1937)},$$

Rips' example showing $F \cap (1 + \mathfrak{f}\mathfrak{s} + \mathfrak{f}^n) \neq S'\gamma_n(F)$ becomes more tantalizing since, in addition, F/S is a finitely generated abelian group. This gives rise to the following problem of independent interest.

Problem. Identify the normal subgroups $F \cap (1 + \mathfrak{f}\mathfrak{s} + \mathfrak{f}^n)$, $n \geq 2$, for F/S finitely generated abelian.

A partial solution of this problem is given by the following theorem which points out the region of possible counter-examples to the dimension subgroup conjecture for metabelian groups for $n \geq 4$.

Theorem. $F \cap (1 + \mathfrak{f}\mathfrak{s} + \mathfrak{f}^n) \leq NS'\gamma_n(F)$, *where* $N = [x_1^{e(1)}, F] \cdots [x_m^{e(m)}, F]$.

Proof. Let w be an arbitrary element of $F \cap (1 + \mathfrak{f}\mathfrak{s} + \mathfrak{f}^n)$. Then $w \in F \cap (1 + \mathfrak{f}^n) = F'$, and it follows that modulo F''' ($\leq S'$), w can be uniquely written as

$$w = \prod_{1 \leq i < j \leq m} [x_i, x_j]^{d(i,j)} = \prod_{1 \leq i \leq m-1} w_i$$

where

$$w_i = \prod_{i+1 \leq j \leq m} [x_i, x_j]^{d(i,j)}$$

and $d(i, j) \in \mathbf{Z}\langle x_i, \cdots x_m \rangle$ (see, for instance, Bachmuth (1965)). Since the ideals $\mathfrak{f}$ and $\mathfrak{s}$ are invariant under the endomorphisms $\theta_k : x_k \mapsto 1, \; x_i \mapsto x_i, \; i \neq k$, we may assume, without loss of generality, that, for each i, $w_i^{-1} \in \mathfrak{fs} + \mathfrak{f}^n$ separately. This yields, in turn,

$$w_i^{-1} \equiv \sum_{i+1 \leq j \leq m} ((x_i - 1)(x_j - 1) - (x_j - 1)(x_i - 1))\, d(i, j) \equiv 0 \bmod \mathfrak{fs} + \mathfrak{f}^n,$$

$$(x_j - 1)(x_i - 1)d(i, j) \equiv 0 \bmod \mathfrak{f}(\mathfrak{s} + \mathfrak{f}^{n-1}),$$

$$(x_i - 1)d(i, j) \equiv 0 \bmod \mathfrak{s} + \mathfrak{f}^{n-1},$$

$$d(i, j) \in ((x_i^{e(i)} - 1)/(x_i - 1))\mathbf{Z}F + \mathfrak{s} + \mathfrak{f}^{n-2}.$$

Since, for $v \in \mathfrak{s} + \mathfrak{f}^{n-2}$, $[x_i, x_j]^v \in [F', S]\gamma_n(F)$, and for $u \in ((x_i^{e(i)} - 1)/(x_i - 1))\mathbf{Z}F$, $[x_i, x_j]^u \in [x_i^{e(i)}, F]$, it follows that $[x_i, x_j]^{d(i,j)} \in NS'\gamma_n(F)$. Thus $w \in NS'\gamma_n(F)$, as required.

Corollary (Higman and Rees). $F \cap (1 + \mathfrak{r} + \mathfrak{f}^3) = R\gamma_3(F)$.

Proof. Since $F \cap (1 + \mathfrak{fr} + \mathfrak{f}^3) \leq F \cap (1 + \mathfrak{fr} + \mathfrak{f}^3) \leq NS'\gamma_3(F)$ and $NS' \leq R\gamma_3(F)$, it follows that $F \cap (1 + \mathfrak{fr} + \mathfrak{f}^3) \leq R\gamma_3(F)$ and the proof follows from Lemma 1.

3. A Theorem of Sjogren and Hartley and Sjogren's Key Lemma for Higher Dimension Subgroups.

The foundation of the following lemma is due to Sjogren (1979) and Hartley (1982). [See Gupta (1987) for details.]

Lemma (Sjogren and Hartley). *Let* $H = H_1 \geq H_2 \geq \cdots$ *and* $K = K_1 \geq K_2 \geq \cdots$ *be series of normal subgroups of a group* F *and let* $\{D_{k,l} : 1 \leq k \leq l\}$ *be a family of normal subgroups of* F *such that*

(a) $D_{k,k+1} = H_k H_{k+1}$;

(b) $H_k H_l \leq D_{k,l}$;

(c) $D_{k,l+1} \leq D_{k,l}$ *for all* $k < l$;

(d) *for each* $2 \leq k + l \leq n - 1$, $k, \; l \geq 1$, *there exsists* $a(k)$ *(depending on* k *and* n*) satisfying* $(K_{k+l} \cap D_{k,k+l+1})^{a(k)} \leq H_k D_{k+1,k+l+1}$.

Then $(D_{1,n})^{a(1,n-1)} \leq H_1 K_n$, *where* $a(1, n-1) = a(1)^{\binom{n-2}{1}} \cdots a(n)^{\binom{n-2}{n-2}}$.

Remark. If we set $H_k = R(k), \; k \geq 1$; $K_l = \gamma_l(F), \; l \geq 1$, and $D_{k,l} = F \cap (1 + \mathfrak{r}(k) + \mathfrak{f}^l), \; 1 \leq k \leq l$, then $\{D_{k,l} : 1 \leq k \leq l\}$ is a family of normal subgroups of F satisfying (a), (b), (c) of the Sjogren and Hartley Lemma. If, in addition,

$$(F \cap (1 + \mathfrak{r}(k) + \mathfrak{f}^{n+2}))^{a(k)} \leq R(k)\gamma_{n+2}(F)(F \cap (1 + \mathfrak{r}(k+1) + \mathfrak{f}^{n+2})),$$

$k = 2, \cdots, n$, for some $a(k) \geq 1$, then condition (d) of the lemma would also be satisfied, and we could conclude that $F \cap (1+\mathfrak{r}+\mathfrak{f}^{n+2})/R\gamma_{n+2}(F)$ has exponent dividing $a(1, n+1) = a(1)^{\binom{n}{1}} \cdots a(n)^{\binom{n}{n}}$.

Sjogren's Key Lemma. *The following inequality holds for all k, $m \geq 1$.*

$$(\gamma_{k+m}(F) \cap (1 + \mathfrak{r}(k) + \mathfrak{f}^{k+m+1}))^{b(k)} \leq R(k)\gamma_{k+m+1}(F)(F \cap (1 + \mathfrak{r}(k+1) + \mathfrak{f}^{k+m+1})),$$

where $b(k) = \text{l.c.m}\{1, \cdots, k\}$.

Thus, using the Sjogren and Hartley Lemma, we obtain Sjogren's Theorem mentioned earlier in the introduction.

Theorem (Sojgren 1979). $F \cap (1+\mathfrak{r}+\mathfrak{f}^{n+2})/R\gamma_{n+2}(F)$ *has exponent dividing* $c_{n+2} = b(1)^{\binom{n}{1}} \cdots b(n)^{\binom{n}{n}}$.

4. Higher Dimension Subgroups of Metabelian Groups.

Let $s = \mathbb{Z}F(F'-1)$. Then $F'' - 1 \leq s^2$. For the proofs of the following theorems, we refer to Gupta (1987)$_b$.

Theorem A (Gupta (1987)$_a$). $(F \cap (1 + \mathfrak{r}(k) + s^2 + \mathfrak{f}^{k+l}))^{d(l)} \leq R(k)F''\gamma_{k+l}(F)$, *where primary decomposition of $d(l)$ uses primes not exceeding l.*

Theorem B. *Sjogren's Key Lemma is valid with $\mathfrak{r}(k)$ replaced by $\mathfrak{r}(k) + s^2$.*

Theorem C (Gupta (1988)). *If G is a finite metabelian p-group, then $D_{n+2}(G) = \gamma_{n+2}(G)$ for all $n \leq 2p - 3$.*

Proof. Choose $a(1) = b(1), \cdots, a(p-1) = b(p-1)$, $a(p) = d(p-1)$, $a(p+1) = d(p-2), \cdots$, and $a(2p-3) = d(2)$. Then p is coprime to each $a(i)$ and the proof follows using Theorem A and B.

We conclude with the following announcement. The details will be published elsewhere.

Announcement. For $k \geq 3$, $n \geq 3$, $F \cap (1 + \mathfrak{r}(k) + s^2 + \mathfrak{f}^n) = R(k)F''\gamma_n(F)$.

Corollary. *If G is a finitely generated metabelian group, then for all $n \geq 2$, $D_n(G)/\gamma_n(G)$ is a 2-group of exponent $2^{\binom{n-2}{2}}$.*

Corollary. *If G is a metabelian p-group, p odd, then $D_n(G) = \gamma_n(G)$ for all $n \geq 2$.*

Cojecture. For each $n \geq 4$, there exists a finite metabelian 2-group G with $D_n(G) \neq \gamma_n(G)$.

References

[1] F. Bachmann and L.Gruenenfelder, 'Homological methods and the third dimension subgroup', *Comment. Math. Helv.* 47, 526-531.

[2] S. Bachmuth (1965), 'Automorphisms of free metabelian groups', *Trans. Amer. Math. Soc.* 118, 93-104.

[3] Otto, Grün (1936), 'Über die Faktorgruppen freier Gruppen I', *Deutsche Math.* (Jahrgang 1), 6, 772-782.

[4] N.D. Gupta, A.W. Hales and I.B.S. Passi (1984), 'Dimension subgroups of metabelian groups', *J. reine angew. Math.* 346, 194-198.

[5] Narain Gupta (1982), 'On the dimension subgroups of metabelian groups', *J. Pure Appl. Algebra* 24, 1-6.

[6] Narain Gupta (1987)a, 'Sjogren's theorem for dimension subgroups – the metabelian case', *Annals of Math. Study* 111, 197-211, Princeton University Press.

[7] Narain Gupta (1987)b, 'Free Group Rings', *Contemporary Math.* 66, Amer. Math. Soc.

[8] Narain Gupta (1988), 'Dimension subgroups of metabelian p-groups', *J. Pure Appl. Algebra* 51, 241-249.

[9] Narain Gupta (1988), 'A theorem of Sjogren and Hartley on dimension subgroups', *Publ. Math. Debrecen*, to appear.

[10] Narain Gupta and Ken-Ichi Tahara (1985), 'Dimension and lower central subgroups of metabelian p-groups', *Nagoya Math. J.* 100, 127-133.

[11] B. Hartley (1982), *Dimension and lower central subgroups – Sjogren's theorem revisited*, Lecture notes 9, National University of Singapore.

[12] Gerald Rosey (1974), 'N-series and the filteration of the augmentation ideal', *Canad. J. Math,* 26, 962-977.

[13] Wilhelm Magnus (1937), 'Über Beziehungen zwischen höheren kommutatoren', J. reine angew. Math. 177, 105-115.

[14] Wilhelm Magnus (1939), 'On a theorem of Marshall Hall', *Ann. Math.* (Ser. II) 40, 764-768.

[15] Wilhelm Magnus, Abraham Karrass and Donald Solitar (1966), *Combinatorial group theory*, Interscience Publ. Co. New York.

[16] I.B.S. Passi (1968), 'Dimension subgroups', *J. Algebra*, 9, 152-182.

[17] Inder Bir S. Passi (1979), *Group rings and their augmentation ideals*, Lecture Notes in Math. 715, Springer-Verlag.

[18] E. Rips, 'On the fourth integer dimension subgroup', *Israel J. Math.* 12, 342-346.

[19] Frank Röhl (1985), 'Review and some critical comments on a paper of Grün concerning the dimension subgroup conjecture', *Pol. Soc. Pras. Mat.* 16, 11-27.

[20] Robert Sandling (1972), 'The dimension subgroup problem', *J. Algebra* 21, 216-231.

[21] Ken-Ichi Tahara (1977), 'On the structure of $Q_3(G)$ and the fourth dimension subgroup', *Japan J. Math.* (N.S.) 3, 381-396.

[22] Ken-Ichi Tahara (1981), 'The augmentation quotients of group rings and the fifth dimension subgroups', *J. Algebra* 71, 141-173.

[23] E. Witt (1937), ' Treue Darstellung Liesche Ringe', *J. reine angew. Math.* 177, 152-160.

Department of Mathematics
University of Manitoba
Winnipeg, R3T 2N2,
Cannada

DOUBLY REGULAR ASYMMETRIC DIGRAPHS WITH RANK 5 AUTOMORPHISM GROUPS

Noboru Ito

1. Introduction.

Let v, k and λ be positive integers. Let $\mathbf{A}$ be the set of all $(0,1)$ matrices A of degree v such that

$$AA^t = (k - \lambda)\mathbf{I} + \lambda\mathbf{J}, \tag{1}$$

where t denotes the transposition, and $\mathbf{I}$ and $\mathbf{J}$ denote the identity and all one matrices of degree v respectively, and that

$$A + A^t \text{ is also a } (0,1) \text{ matrix.} \tag{2}$$

Two matrices A_1 and A_2 of $\mathbf{A}$ are equivalent if there exists a permutation matrix P such that $A_2 = P^{-1}A_1P$.

Now we regard A as the adjacency matrix of a digraph $D = (V, \mathbf{A})$, where V and $\mathbf{A}$ denote the sets of vertices and arcs of D respectively, defined as follows: $V = \{1, 2, \cdots, v\}$ and $(i, j) \in \mathbf{A}$ if and only if $A(i, j) = 1$, where $A(i, j)$ is the (i, j) component of A, $1 \leq i, j \leq v$.

(1) shows that A may be regarded as an incidence matrix of a symmetric $2\text{-}(v, k, \lambda)$ design. This fact implies that D is regular of valency k and doubly regular of double valency λ (see [4]). Furthermore (2) shows the asymmetry of D. So we call D a *doubly regular asymmetric* (v, k, λ) *digraph*.

The set of all permutation matrices P such that $P^{-1}AP = A$ forms a subgroup G of the symmetric group $\mathrm{Sym}(v)$ which is consisted of all permutation matrices of degree v. G is isomorphic to the automorphism group of D.

In this paper we assume that G is of rank 5 as a permutation group on V. It is rather obvious (see Section 2) that a one point stabilizer of G has at least one pair of paired orbits. Now the main result is the following theorem.

Theorem. *If a one point stabilizer of G has two pairs of paired orbits, then A is equivalent to the cyclic incidence matrix of the cyclic difference set in $GF(p)$, where $v = p = 4x^2 + 1$ with x odd ≥ 3 is a prime, consisting of all biquadratic residues.*

There are some partial results in the case where a one point stabilizer of G has only one pair of paired orbits.

2. Preliminaries on automorphism groups

Let $D = (V, A)$ be a doubly regular asymmetric (v, k, λ) digraph, where V and A denote the sets of vertices and arcs of D respectively. Let G be the automorphism group of D.

Proposition 1. *Let a be a vertex of D, $D_1(a)$ the out–neighborhood of a and $D_{-1}(a)$ the in–neighborhood of a. Let $G(a)$ be the stabilizer of a in G. Then $G(a)$ has the same number of orbits on $D_1(a)$ and on $D_{-1}(a)$.*

Proof. An orbit of $G(a)$ corresponds to a double coset of $G(a)$. Let $G(a)\sigma_1 G(a)$, $G(a)\sigma_2 G(a)$, $\cdots$, $G(a)\sigma_r G(a)$ be the set of double cosets of $G(a)$ corresponding to the decomposition of $D_1(a)$ into orbits of $G(a)$. Then $G(a)\sigma_1^{-1}G(a), G(a)\sigma_2^{-1}G(a), \cdots, G(a)\sigma_r^{-1}G(a)$ form the set of double cosets of $G(a)$ corresponding to the decomposition of $D_{-1}(a)$ into orbits of $G(a)$.

Let $N(a)$ be the set of vertices c of D such that c and a are non–adjacent. Now we assume that G has rank 5 on V.

Proposition 2. $G(a)$ *is transitive on $D_1(a)$ and on $D_{-1}(a)$. $N(a)$ decomposes into two orbits, say $N_1(a)$ and $N_2(a)$.*

Proof. By Proposition 1, otherwise, $G(a)$ has two orbits on $D_1(a)$ and on $D_{-1}(a)$, and we have that $V = \{a\} + D_1(a) + D_{-1}(a)$. So D is a Hadamard tournament and hence G has odd order, which contradicts the fact that $G(a)$ has two orbits on $D_1(a)$.

From the proof of Proposition 1 we see that $D_1(a)$ and $D_{-1}(a)$ are paired orbits (See [10], p.44). We distinguish two cases: Case where $N_1(a)$ and $N_2(a)$ are paired orbits and Case where $N_1(a)$ and $N_2(a)$ are self-paired orbits.

3. Paired case.

In this section we assume that $N_1(a)$ and $N_2(a)$ are paired orbits of $G(a)$.

Lemma. *If $v - 2k$ divides v, then $v = 4(k - \lambda)$.*

Proof. If $v - 2k$ divides v, put $v = m(v - 2k)$ with m a positive integer. Hence, $(m-1)v = 2mk$. So we may put $v = mw$ with w a positive integer. Hence, $(m-1)w = 2k$. Since $\lambda(v-1) = k(k-1)$, we have that $4\lambda(mw - 1) = (m-1)w\{(m-1)w - 2\}$. So we may put $4\lambda = xw$ with x a positive integer. Hence, $x = m - 1 - \{(m-1)(w+1)/(mw-1)\}$. If $(m-1)(w+1) = mw - 1$, then $m = w$ and $x = m - 2$, which implies that $v = 4(k - \lambda)$. If $(m-1)(w+1) \geq 2mw - 2$, then $w = 1$, $v = m$ and $v = 2k + 1$ against the rank 5 assumption.

Now by ([10], (16.5)) G is of odd order. Therefore by the Feit-Thompson theorem G is solvable.

Proposition 3. *G is primitive on V.*

Proof. Assume not and let M be a maximal subgroup of G containing $G(a)$. Since D is strongly connected (See [4]), the orbit D of M containing the vertex a is disjoint with $D_1(a)$ and $D_{-1}(a)$. Since the order of G is odd, we have that $D = \{a\} + N(a)$. Thus the index of $G(a)$ in M equals $v - 2k$ contradicting Lemma. So $v = p^a$, where p is a prime and a is a positive integer. Moreover G contains a regular normal subgroup P which is an elementary Abelian p–group; $G = PG(a)$ and $P \cap G(a) = \{e\}$.

Proposition 4. *G is $(3/2)$-transitive on V.*

Proof. We may regard P as an a–dimensional vector space over $\mathrm{GF}(p)$ and $G(a)$ as a linear group over P, where the action of elements of $G(a)$ corresponds to a conjugation.

First assume that $P \equiv 1 \pmod 4$. Then the multiplicative group $\mathrm{GF}(p)^e$ contains an element i of order 4 and $i\mathrm{I}_a$, where I_a is the identity operator on P, commutes with every element of $G(a)$ and has no fixed non–zero vector. So $G(a)$ has at least four orbits of the same length on P. This implies that G is $(3/2)$–transitive on V.

Secondly assume that $p \equiv 3 \pmod 4$. Put $|N_1(a)| = |N_2(a)| = l$. Then $v = p^a = 1+2k+2l$. Since k and l are odd, $v \equiv 1 \pmod 4$. So a is even; $a = 2b$ with b a positive integer. Hence $v \equiv 1 \pmod 8$ and $k + l \equiv 0 \pmod 4$. Since k and l are odd, $k \not\equiv l \pmod 4$. On the other hand, by the Fong-Swan theorem ([2], p.419) P is not absolutely irreducible. In fact, over $\mathrm{GF}(p^2)$ $G(a)$ decomposes into two algebraically conjugate parts; $G(a) = G_1(a) \oplus G_2(a)$, where $G_2(a)$ is obtained from $G_1(a)$ by a field automorphism $x \mapsto x^p$, $x \in \mathrm{GF}(p^2)$. Let $P \otimes \mathrm{GF}(p^2) = P_1 \oplus P_2$ be the corresponding decomposition of $P \otimes \mathrm{GF}(p^2)$ and I_b the identity operator on P_i $(i = 1, 2)$. Further let i be an element of order 4 of the multiplicative group $\mathrm{GF}(p^2)^x$. Then $i\mathrm{I}_b + (-i)\mathrm{I}_b$ commutes with every element of $G_1(a) \oplus G_2(a)$. Therefore there exists a linear operator of order 4 of P which commutes with every element of $G(a)$ and has no non–zero fixed vector. So $G(a)$ has at least four orbits of the same length on P. This implies a contradiction that $k = l$.

In particular, we have shown that $p \equiv 1 \pmod 4$. Since G is $(3/2)$–transitive and of rank 5 on V, $v = p^a = 4k + 1$ and $k = 4l + 1$. Since k is odd, a is odd.

Proposition 5. *We may assume that $\cdot P = \mathrm{GF}(p^a)$ and that $G(a)$ is contained in the group of semilinear transformations : $x \mapsto ax^\sigma + b$, where $a \neq 0$, $b \in \mathrm{GF}(p^a)$ and σ is an automorphism of $\mathrm{GF}(p^a)$.*

Proof. We may assume that $a > 1$. Moreover, by a result of Passman ([7], Theorem B), we may assume that G is a Frobenius group. Now there exists a prime factor r of $p^a - 1$ such that the order of $r \pmod p$ equals a ([8]). Let R be a Sylow r–subgroup of $G(a)$. Then R is cyclic and non–trivial. We show that R is normal in $G(a)$. Deny. Then there exists an s–subgroup S of $G(a)$, where s is a prime, such that S is a normal subgroup of a non–Abelian subgroup RS. Now consider P as a representation module of RS. Then since S is cyclic and since rs prime to p, we have that $a \equiv 0 \pmod r$ ([1], (9.13)). Since $r \equiv 1 \pmod a$, this is a contradiction. So by a result of Huppert ([3], Hilfssatz 2) we get Proposition 5.

So we may assume that D comes from the difference set in $GF(p^a)$ consisting of biquadratic residues (= fourth powers). Moreover, we have that $a = 1$ and that k is a square ([6], Theorem 8.6 and a remark after it). Thus we obtain the theorem in Section 1 (See also [5], Remark to Proposition 24).

We add the following proposition concerning the structure of D.

Proposition 6. *Assume that* $2\mu_1 = 2\lambda + 1 + x$, $2\mu_2 = 2\lambda + 1 - x$ *and* $x^2 = k$. *Then*

$$A^2 = \lambda A + \lambda A^t + \mu_1 N_1 + \mu_2 N_2 \qquad (3)$$

Moreover, every diagonal component of A^3 *equals* $k\lambda$. *In particular,* D *has diameter 2 and girth 3.*

Proof. We adjust the adjacency matrix A of D so that the first row corresponds to $0 \in GF(p)$ and that $A(1,i) = 1$ if and only if $i - 1 \in GF(p)$ is a biquadratic residue and that A is cyclic. Then $A^2(1,i)$ is equal to the number $n(h,j)$ of pairs (h,j) such that $(h-1)+(j-1) = i-1$, where $h - 1$ and $j - 1$ are biquadratic residues. Then, since $(h-1)(j-1)^{-1} + 1 = (i-1)(j-1)^{-1}$, we see that $n(h,j)$ equals a cyclotomic number. So (3) follows from ([9], Lemma 19). Since trace(A^2) = trace(AN_1) = trace(AN_2) = 0 and $AA^t = (k - \lambda)I + \lambda J$, we get the second assertion from (3).

Remark. There seem to exist a huge number of doubly regular asymmetric digraphs associated with a given difference set of a biquadratic residue type. For $p = 37$, $D_1(0) = \{1, 7, 9, 10, 12, 16, 26, 33, 34\}$ and $D_1(i) = \{1+i, 7+i, 9+i, 10+i, 12+i, 16+i, 26+i, 33+i, 34+i\}$, where numbers are reduced mod 37 and $1 \leq i \leq 36$, define our rank 5 digraph. But we can define a doubly regular asymmetric (37,9,2) digraph D'' as follows, where $D_1''(i)$ denote the out–neighborhood of the vertex i in D'': $D_1''(i) = D_1(i)$ for $i = 0, 1, \cdots, 16, 22, 23, 26, 27, 30, 31$. $D_1''(17) = D_1(25)$, $D_1''(18) = D_1(24)$, $D_1''(19) = D_1(17)$, $D_1''(20) = D_1(28)$, $D_1''(21) = D_1(19)$, $D_1''(24) = D_1(21)$, $D_1''(25) = D_1(33)$, $D_1''(28) = D_1(36)$, $D_1''(29) = D_1(35)$, $D_1''(32) = D_1(29)$, $D_1''(33) = D_1(20)$, $D_1''(34) = D_1(32)$, $D_1''(35) = D_1(18)$, $D_1''(36) = D_1(34)$. It is easy to see that D'' has diameter 3 and that the automorphism group of D'' is trivial.

4. Self-paired case.

It is not known to us whether this case really occurs or not. In this paper we only show that A has an eigenvalue r which is a rational integer such that $k - \lambda = r^2$.

We may use vertices of D as row and column labels of various matrices of degree v. So if A is the adjacency matrix of D, $A(a,b) = 1$ if and only if (a,b) is an arc of D. Moreover let N_i be the matrix of degree v such that $N_i(a,b) = 1$ if and only if b belongs to $N_i(a)$, where $i = 1,2$. Then we have that

$$A + A^t + I + N_1 + N_2 = J. \qquad (4)$$

In the present case N_1 and N_2 are symmetric matrices. We may regard G as the group of permutation matrices of degree v. Since G is of rank 5, we may assume that G is decomposed into the following form:

$$G = \text{diagonal}\,(1_G,\ X_2,\ X_3,\ X_4,\ X_5), \tag{5}$$

where 1_G denotes the identity representation of G and the X_i $(i = 2, 3, 4, 5)$ are distinct irreducible non–identity components of G. So the centralizer ring C of G is commutative and $\{A, A^t, I, N_1, N_2\}$ is a basis for C ([10], p.86). We may bring matrices in (4) into diagonal forms according to the decomposition (5). In addition, corresponding eigenvalues of A and A^t are complex-conjugate. Moreover, since A is not symmetric, A has a non–real eigenvalue.

Proposition 7. *A has a real eigenvalue other than k.*

Proof. Put

$$A^2 = \lambda A + xA^t + y_1 N_1 + y_2 N_2, \tag{6}$$

where x, y_1 and y_2 are non–negative integers. The fact that the coefficient of A equals λ comes from the fact that the out–degree of vertices of $D_1(a)$ within $D_1(a)$ equals λ and since $G(a)$ is transitive on $D_1(a)$, the out–degree equals the in–degree, where $a \in V$.

Now assume that A has no real eigenvalue other than k, and let $a + bi$ be a non–real eigenvalue of A. Then by (6) we obtain that

$$(a + bi)^2 = \lambda(a + bi) + x(a - bi) + y_1 m_1 + y_2 m_2, \tag{7}$$

where m_1 and m_2 are (real) eigenvalues of N_1 and N_2 respectively corresponding to $a + bi$. Comparing the imaginary parts of both sides of (7) we have that $2ab = (\lambda - x)b$. Since b is not equal to zero, we have that

$$2a = \lambda - x. \tag{8}$$

Now, since $\text{trace}(A) = 0$, we obtain that $2k + (v - 1)(\lambda - x) = 0$. Since $v - 1 > 2k$, this is a contradiction (See [4]).

Since $AA^t = (k - \lambda)I + \lambda J$, the square of a real eigenvalue in Proposition 7 is equal to $k - \lambda$.

Proposition 8. *$k - \lambda$ is the square of a rational integer.*

Proof. Assume that $k - \lambda$ is not a square. Then both $(k - \lambda)^{1/2}$ and $-(k - \lambda)^{1/2}$ are eigenvalues of A with the same multiplicity. So spectra of A and A^2 are as follows:

$$\text{Spec}(A) = \begin{pmatrix} k & (k-\lambda)^{1/2} & -(k-\lambda)^{1/2} & a+bi & a-bi \\ 1 & f_2 & f_2 & f_4 & f_4 \end{pmatrix}$$

and

$$\cdot\,\text{Spec}(A^2) = \begin{pmatrix} k^2 & k-\lambda & a^2-b^2+2abi & a^2-b^2-2abi \\ 1 & 2f_2 & f_4 & f_4 \end{pmatrix},$$

where $f_2 = deg(X_2) = deg(X_3)$ and $f_4 = deg(X_4) = deg(X_5)$. Since $\text{trace}(A) = \text{trace}(A^2) = 0$, we obtain that

$$k + 2af_4 = 0 \tag{9}$$

and

$$k^2 + 2f_2(k - \lambda) + 2f_4(a^2 - b^2) = 0. \tag{10}$$

By (8) and (9) we see that f_4 divides k. Since $a^2 + b^2 = k - \lambda$, from (10) it follows that $k^2 + 2f_2(k-\lambda) = 2f_4(k - \lambda - 2a^2)$, which implies that $2f_4 > k$. Thus we obtain $f_4 = k$ and $2a = -1$. So from (10) it follows that

$$2f_2(k - \lambda) = k^2 - 2\lambda k - k. \tag{11}$$

On the other hand, from $v = 2k + 2f_2 + 1$ and $\lambda(v - 1) = k(k - 1)$ it follows that

$$2f_2\lambda = k^2 - 2\lambda k - k. \tag{12}$$

From (11) and (12) we get $k = 2\lambda$, which leads to a contradiction $v = 2k - 1$ (See [4]).

References

[1] Feit, W. *Characters of finite groups*, Benjamin, 1967.

[2] Feit, W. *The representation theory of finite groups*, North-Holland, 1982.

[3] Huppert, B. 'Zweifach transitive, auflösbare Permutationsgruppen', *Math. Zeitschr.*, 68(1957), 126-150.

[4] Ito, N. 'Doubly regular asymmetric digraphs', to appear in *Discrete Math.*, 72 (1988).

[5] Ito, N. 'Automorphism groups of DRADs', to appear in *Proc. Singapore Group Theory Conference*.

[6] Mann, H.B. *Addition theorems*, Krieger, 1976.

[7] Passman, D.S. 'Exceptional 3/2–transitive permutation groups', *Pacific J. Math.*, 29(1969), 669-713.

[8] Rédei, L. 'Über die algebraischzahlentheoretische Verallgemeinerung eines elementarzahlentheoretischen Satzes von Zsigmondy', *Acta Sci. Math.*, Szeged 19(1958), 98-126.

[9] Storer, T. *Cyclotomy and difference sets*, Markham, 1967.

[10] Wielandt, H. *Finite permutation groups*, Academic Press, 1964.

Department of Applied Mathematics
Konan University
Kobe 658
Japan

EXTENSIONS OF THE CLASS OF LIE ALGEBRAS WITH THE LATTICE OF SUBIDEALS

Naoki Kawamoto and N. Nomura

0. Introduction.

Interesting properties are known about subideals of not necessarily finite dimentional Lie algebras. In this paper we consider the class $\mathfrak{L}$ of Lie algebras in which the join of any two subideals is a subideal. Honda [2,3] has considered the class $\mathfrak{L}$ and shown that several classes are contained in $\mathfrak{L}$. Let $\mathfrak{X}$ be a class of Lie algebras, which is closed for taking subideals and quotient algebras, and suppose that abelian Lie algebras in the class $\mathfrak{X}$ are finite dimensional. We shall show that the classes $\mathfrak{X} \mathfrak{L}$ and $\mathfrak{L} \mathfrak{X}$ are contained in $\mathfrak{L}$ over a field of characteristic zero (Theorems 1 and 2). Robinson [5,6], Smith [7], and Lennox and Stonehewer [4] have shown interesting properties of a similar class of groups, and we are inspired by their results.

1. Notations and Preliminaries.

We suppose that the ground field $\mathfrak{k}$ is of characteristic zero, and that Lie algebras are not necessarily finite dimensional over $\mathfrak{k}$. Over a field of positive characteristic Hartley (see [1, Lemma 3.1.1]) has given an example of Lie algebra which is finite dimensional and soluble but not belongs to $\mathfrak{L}$, and hence we restrict ourselves to a field of characteristic zero.

We write $H \lhd L$ when H is an ideal of a Lie algebra L. A subalgebra H of L is called a *subideal* of L if there exists an integer n and a chain

$$H = H_0 \lhd \cdots \lhd H_i \lhd H_{i+1} \lhd \cdots \lhd H_n = L,$$

of subalgebras H_i of L. In this case, We write H si L or $H \lhd^n L$.

We use the class notations: $\mathfrak{F}$, $\mathfrak{A}$, $\mathfrak{N}$, and $E\mathfrak{A}$ are the classes of finite dimensional, abelian, nilpotent, and soluble Lie algebras respectively. $\mathfrak{D}$ and $\mathfrak{T}$ are the classes of Lie algebras in which every subalgebra is a subideal and every subideal is an ideal, respectively. Max-si (resp. Min-si) is the class of Lie algebras which satisfies the maximal (resp. minimal) condition for subideals. $\mathfrak{G}^I$ is the class of Lie algebras whose subideals are finitely generated. If $\mathfrak{X}$ and $\mathfrak{Y}$ are classes of Lie algebras then $\mathfrak{X} \mathfrak{Y}$ denotes the class of Lie algebras L with an ideal $I \in \mathfrak{X}$ and a quotient algebra $L/I \in \mathfrak{Y}$. It is clear that $\mathfrak{X}(\mathfrak{Y}\mathfrak{Z}) \le (\mathfrak{X}\mathfrak{Y})\mathfrak{Z}$, and we simply write $\mathfrak{X} \mathfrak{Y} \mathfrak{Z}$ for $(\mathfrak{X} \mathfrak{Y}) \mathfrak{Z}$. A Lie algebra L is an $\mathfrak{X}$-algebra if L belongs to a class $\mathfrak{X}$. A class $\mathfrak{X}$ is called I-closed (resp. Q-closed) if any subideal (resp. any quotient algebra) of an $\mathfrak{X}$-algebra belongs to $\mathfrak{X}$. $\mathfrak{X}$ is $\{I,Q\}$-closed if $\mathfrak{X}$ is I-closed and Q-closed.

It is clear that $\mathfrak{A} \leq \mathfrak{R} \leq \mathfrak{D} \leq \mathfrak{L}$ and $\mathfrak{T} \leq \mathfrak{L}$. A class of Lie algebras is said to be coalescent if the join of any two $\mathfrak{X}$-subideals of a Lie algebra L is an $\mathfrak{X}$-subideal of L for any Lie algebra L. It is well-known that classes $\mathfrak{F}$, Max-si, Min-si, and $\mathfrak{G}^I$ are coalescent over a field of characteristic zero (see [1, Theorems 3.2.5 and 3.3.3]). It is easy to see that $\mathfrak{X} \leq \mathfrak{L}$ whenever $\mathfrak{X}$ is I-closed and coalescent. Hence $\mathfrak{F}$, Max-si, Min-si, and $\mathfrak{G}^I$ are contained in $\mathfrak{L}$. Honda [2,3] has shown that $\mathfrak{D}\,\mathfrak{A}$, $\mathfrak{R}\,\mathfrak{T}$, $\mathfrak{R}\,\mathfrak{X}$, where $\mathfrak{X}$ is I-closed and coalescent, and several other classes are contained in $\mathfrak{L}$.

Let A and B be subspaces of a Lie algebra L. We denote by A^B the minimum subspace of L which contains A and is invariant by ad x ($x \in B$). If ad x is nilpotent for an element $x \in L$, then $\exp\,\text{ad}\,x = \sum_{n=0}^{\infty}(1/n!)(\text{ad}\,x)^n$ is an automorphism of L [1, Theorem 1.4.7]. A^θ denotes the image of A under an automorphism θ of L.

Subspaces A and B of a Lie algebra L are called *permutable* if $[A,B] \subseteq A+B$. The *permutaiser* P of A in B is the sum of all subspaces of B which permutes with A. Thus P is the largest subspace of B which permutes with A. The other notations and terminology not mentioned here may be found in [1].

2. Soluble $\mathfrak{L}$-algebras.

We begin with the following lemmas but we omit the proofs.

Lemma 1. ([1], Lemma 2.2.3). *Let A and B be permutable subspaces of a Lie algebra L, and M be a subspace of L. Then*

$$M^{\langle A,B\rangle} = M^{A+B} = M^{A\cup B} = (M^A)^B = (M^B)^A.$$

Lemma 2. ([1], Lemma 2.1.4). *Let H and K be permutable subideals of a Lie algebra L, Then $\langle H,K\rangle$ is a subideal of L*

Lemma 3. ([2], Lemma 2.4). *Let H and K be subideals of a Lie algebra L. Then $\langle A,B\rangle$ si L if and only if $\langle H^K\rangle$ si L.*

Proof. Assume that $\langle A,B\rangle$ si of L. Then $\langle H^K\rangle$ si L since $\langle H^K\rangle \lhd \langle H,K\rangle$. Conversely suppose that $\langle H^K\rangle$ si L. Then by Lemma 2 $\langle H,K\rangle = \langle H^K\rangle + K$ si L since $\langle H^K\rangle$ and K are permutable.

We now show the following result which has a counterpart ([4], Proposition 3.5.2) in group theory.

Proposition 1. *Let $\mathfrak{X}$ be an $\{I,Q\}$-clodsed class of Lie algebras. Then $\mathfrak{X} \leq \mathfrak{L}$ if and only if*

$\mathfrak{X} \cap E\mathfrak{A} \leq \mathfrak{L}$.

Proof. Assume that $\mathfrak{X} \cap E\mathfrak{A} \leq \mathfrak{L}$. Let $L \in \mathfrak{X}$, $H \lhd^n L$, $K \lhd^m L$, and $J = \langle H, K \rangle$. We show that J si L by induction on n. The case $n = 1$ is clear. Assume that $n > 1$ and that the result holds for $n - 1$. Let $H \lhd H_1 \lhd^{n-1} L$, and $M = \langle H_1, K \rangle$. Then M si L. Let P be the permutiser of K in H_1. Then P si L by [1, Lemma 2.1.5]. Since P and K are permutable, $P + K$ si L by Lemma 2. Since H is invariant by ad x for $x \in H_1$, we have $H^{P+K} = H^K$ by Lemma 1. If $N = \langle H, P + K \rangle$, then $\langle H^K \rangle = \langle H^{P+K} \rangle \lhd N$, and K si N. Hence $J = \langle H^K \rangle + K$ si N. By [1, Corollary 2.2.5] $H_1^{(q)} \subseteq P$ for some integer q, whence by [1, Theorem 2.2.7] $M^{(r)} \subseteq H_1^{(q)} + K$ for some integer r. Therefore $M^{(r)} \subseteq P + K$. Since M si L and $\mathfrak{X}$ is $\{I,Q\}$-closed, it follows that $M/M^{(r)} \in \mathfrak{X} \cap E\mathfrak{A} \leq \mathfrak{L}$. Now $H + M^{(r)}/M^{(r)}$ and $P + K/M^{(r)}$ are subideals of $M/M^{(r)}$, whence $N/M^{(r)}$ si $M/M^{(r)}$. Thus N si M, and we have J si N si M si L.

3. Invariant subalgebras.

We give a generalization of [1, Lemma 1.4.10] in the following.

Lemma 4. *Let M be a subspace of a Lie algebra L, and let X be a finitely generated subalgebra of L. If $[L_{,n} X] = 0$ for some integer n, then $M^X = M^{\theta_1} + \cdots + M^{\theta_m}$ for some $\theta_1, \cdots, \theta_m \in \langle \exp \text{ad } x \mid x \in X \rangle$.*

Proof. Let $X = \langle x_1, \cdots, x_r \rangle$. Then

$$M^x = \sum_{i=0}^{n-1} [M_{,i} X] = \sum_{i=0}^{n-1} \sum_{1 \leq i_1, \cdots, i_i \leq r} [M, x_{i_1}, \cdots, x_{i_i}].$$

If we put $\delta = \text{ad x}$, then

$$y^{\exp(i\delta)} = y + iy\delta + \cdots + (i^{n-1}/(n-1)!)y\delta^{n-1}$$

for any $y \in M$, and $i \in \mathbf{N}$. The matrix $A = (a_{ij})$ with $a_{ij} = i^j/j!$ $(1 \leq i \leq n, 0 \leq j \leq n-1)$ is non-singular, and hence there exist $\alpha_1, \cdots \alpha_n \in \mathbf{t}$ such that $\sum_i \alpha_i a_{ij} = \delta_{ij}$. Now $y\delta = \sum_i \alpha_i y^{\exp i\delta}$ and $[M, x] \subseteq \sum_i M^{\exp i\delta}$. Therefore $[M, x_j] \subseteq \sum_i M^{\theta_i}$, where $\theta_i = exp(i \text{ ad } x_j)$. Inductively there exists an integer k satisfying

$$[M, x_{i_1}, \cdots, x_{i_i}] \subseteq \sum_{i=0}^{k} M^{\theta_i}, \quad \theta_i \in \langle \exp \text{ad } x \mid x \in X \rangle.$$

Hence $M^X \subseteq \sum_{i=1}^{m} M^{\theta_i}$ for some m and we have $M^X = \sum_{i=1}^{m} M^{\theta_i}$ for some m.

Lemma 5. *Let H and K be subalgebras of a Lie algebra L. Suppose that $K = X + N$, where X is a finitely generated subalgebra of K and N is an ideal of K, and let $[L_m X] = 0$ for some integer n. Then $\langle H^K \rangle = \langle \langle H^{\theta_1}, \cdots, H^{\theta_m} \rangle^N \rangle$ for some $\theta_1, \cdots, \theta_m \in \langle \exp \text{ad } x \mid x \in X \rangle$.*

Proof. By Lemma 1, $H^K = (H^X)^N$ since X and N are permutable. Now by Lemma 4, $H^X = \sum_{i=1}^{m} H^{\theta_i}$ for some $\theta_1, \cdots, \theta_m \in \langle \exp \text{ad } x \mid x \in X \rangle$, and therefore $\langle H^K \rangle = \langle \langle H^{\theta_1}, \cdots, H^{\theta_m} \rangle^N \rangle$.

Proposition 2. *Let A and N be ideals of a Lie algebra L, and let H and K be subideals of L. Suppose that $L = \langle H, K \rangle + A$, $A \subseteq N$, $A \in \mathfrak{A}$, $L/A \in \mathfrak{L}$, $N \in \mathfrak{L}$, and $L/N \in \mathfrak{F}$. Then $\langle H, K \rangle$ si L.*

Proof. Let $J = \langle H, K \rangle$, $H \lhd^n L$, and $K \lhd^m L$. We use induction on n. If $n = 1$ then clearly $J = H + K$ si L. Let $n > 1$. Put $C = \{ x \in J \mid [x, A] \subseteq A \cap J \}$. It is easy to see that $C \lhd J$, and therefore $C \lhd L$. Since $[A, K^m] \subseteq [A, _m K] \subseteq A \cap K$, we have $K^m \subseteq C$. Hence $[L, _m K, _{m-1} K] \subseteq [K, _{m-1} K] \subseteq C$. We now consider subalgebras in L/C and by abuse of notations we use the same notations to show the subalgebras of L and their homomorphic images in L/C. Thus $H \lhd^n L$, $K \lhd^m L$, and $[L, _{2m-1} K] = 0$. Since $K \cap N \lhd K$ and $K/K \cap N \simeq K + N/N \subseteq L/N \in \mathfrak{F}$, there exists a finitely generated subalgebra X of K satisfying $K = X + K \cap N$. Then by Lemma 5

$$\langle H^K \rangle = \langle \langle H^{\theta_1}, \cdots, H^{\theta_s} \rangle^{K \cap N} \rangle$$

for some $\theta_1, \cdots, \theta_s \in \langle \exp \operatorname{ad} x \mid x \in X \rangle$. Let $H_1 = H^L$. Then $H^{\theta_t} \lhd^{n-1} H_i^{\theta_t} = H_1 \lhd L$ $(1 \le t \le s)$. Put $M = \langle H^{\theta_1}, \cdots, H^{\theta_s} \rangle$. We claim that M si H_1 by induction on s. Let $M_t = \langle H^{\theta_1}, \cdots, H^{\theta_t} \rangle$ $(1 \le t \le s)$. Clearly M_1 si H_1. Assume that $t > 1$ and that M_{t-1} si H_1. Then $M_t + A/A = \langle M_{t-1} + A/A, H^{\theta_t} + A/A \rangle$ si L/A since $L/A \in \quad$, and hence $M_t + A$ si L. Put $L_t = H_1 \cap (M_t + A)$. Then L_t si L, $L_t = M_t + H_1 \cap A$, and $L_t \cap A = H_1 \cap A$. Thus $L_t = \langle M_{t-1}, H^{\theta_t} \rangle + L_t \cap A$. Since L_t si L, it is not hard to see that ideals $L_t \cap A$, $L_t \cap N$ of L_t satisfy the conditions of the proposition. Now $H^{\theta_t} \lhd^{n-1} L_t$, and therefore $M_t = \langle M_{t-1}, H^{\theta_t} \rangle$ si L_t by the inductive hypothesis on n. It follows that M_t si L, and hence M_t si H_1, which completes the second induction. Now we have $M \lhd^k H_1 \lhd L$ for some k. Since $K \cap N$ si L the following conditions are equivalent by Lemma 3: (a) J si L, (b) $\langle H^K \rangle = \langle M^{K \cap H} \rangle$ si L, (c) $\langle M, K \cap N \rangle$ si L, (d) $\langle (K \cap N)^M \rangle$ si L. Hence it is sufficient to show the condition (d). Now there exists a finitely generated subalgebra Y of M such that $M = Y + M \cap N$, since $M/M \cap N \simeq M + N/N \le L/N \in \mathfrak{F}$. By the argument similar to the above we may assume that $[L, _{2(k+1)-1} M] = 0$. Then by Lemma 5 we have

$$\langle (K \cap N)^M \rangle = \langle \langle (K \cap N)^{\tau_1}, \cdots, (K \cap N)^{\tau_t} \rangle^{M \cap N} \rangle$$

for some $\tau_1, \cdots, \tau_t \in \langle \exp \operatorname{ad} y \mid y \in Y \rangle$. Since $K \cap N$ si N, $(K \cap N)^{\tau_i}$ si $N^{\tau_i} = N \in \mathfrak{L}$. Thus T si N, where $T = \langle (K \cap N)^{\tau_1}, \cdots, (K \cap N)^{\tau_t} \rangle$. Since $M \cap N$ si N and $N \in \mathfrak{L}$, we have $\langle (M \cap N), T \rangle$ si $N \lhd L$. Therefore $\langle (K \cap N)^m \rangle = \langle T^{M \cap N} \rangle$ si L, and the proof is completed.

4. Extensions of the class $\mathfrak{L}$.

Let $\mathfrak{X}$ be an $\{I, Q\}$-closed class of Lie algebras. Then clearly $\mathfrak{X} \cap E\mathfrak{A} \le \mathfrak{F}$ if and only if $\mathfrak{X} \cap \mathfrak{A} \le \mathfrak{F}$. In this case $\mathfrak{X} \le \mathfrak{L}$ by Proposition 1 since $\mathfrak{F} \le \mathfrak{L}$. It is known that $\mathfrak{L}\mathfrak{L}$ is not contained in $\mathfrak{L}$ by an example of Hartley (see [1], Lemma 2.1.11), so we have the following.

Theorem 1. *Let $\mathfrak{X}$ be an $\{I, Q\}$-closed class. If $\mathfrak{X} \cap \mathfrak{A} \le \mathfrak{F}$, then $\mathfrak{L}\mathfrak{X} \le \mathfrak{L}$.*

Proof. It is clear that $\mathfrak{L}\mathfrak{X}$ is $\{I, Q\}$-closed. We shall show that $(\mathfrak{L} \cap E\mathfrak{A})(\mathfrak{F} \cap E\mathfrak{A}) \le \mathfrak{L}$, then by Proposition 1 and by the relation $(\mathfrak{L}\mathfrak{X}) \cap E\mathfrak{A} \le (\mathfrak{L} \cap E\mathfrak{A})(\mathfrak{F} \cap E\mathfrak{A})$ we have $\mathfrak{L}\mathfrak{X} \le \mathfrak{L}$. Let $L \in (\mathfrak{L} \cap E\mathfrak{A})(\mathfrak{F} \cap E\mathfrak{A})$, $I \lhd L$, $I \in \mathfrak{L} \cap E\mathfrak{A}$, and $L/I \in \mathfrak{F} \cap E\mathfrak{A}$. Let H and K be subideals of L and $J = \langle H, K \rangle$. Assume that $I^{(r)} = 0$. We show that J si L by induction on r. If $r = 0$ then

$L \in \mathfrak{F} \cap E\mathfrak{A}$ and J si L by the coalescence of $\mathfrak{F} \cap E\mathfrak{A}$. Assume that $r > 0$. If $A = I^{(r-1)}$, then $L/A \in \mathfrak{L}$ by the inductive hypothesis. Let $M = J + A$. Then $M/A = \langle H + A/A, K + A/A \rangle$ si L/A and M si L. Put $N = I \cap M$. Then $A \subseteq N \lhd M$, and $N \in \mathfrak{L}$ since N si I. Now $M/N \in \mathfrak{F}$ since $M/N \simeq M + I/I \leq L/I$. By Proposition 2 we have $\langle H, K \rangle$ si M, and hence $\langle H, K \rangle$ si L.

Theorem 2. *Let $\mathfrak{X}$ be an $\{I,Q\}$-closed class. If $\mathfrak{X} \cap \mathfrak{A} \leq \mathfrak{F}$, then $\mathfrak{X}\mathfrak{L} \leq \mathfrak{L}$.*

Proof. By Proposition 1 it is sufficient to show that $(\mathfrak{X} \cap \mathfrak{L}) \cap E\mathfrak{A} = (\mathfrak{X} \cap E\mathfrak{A})(\mathfrak{L} \cap E\mathfrak{A}) \leq \mathfrak{L}$. Let $L \in (\mathfrak{X} \cap E\mathfrak{A})(\mathfrak{L} \cap E\mathfrak{A})$, $I \lhd L$, $I \in \mathfrak{X} \cap E\mathfrak{A}$, and $L/I \in \mathfrak{L} \cap E\mathfrak{A}$. Suppose that H and K be subideals of L and $J = \langle H, K \rangle$. Then $J + I/I$ si L/I and $J + I$ si L. Let $f : J \longrightarrow \text{Der}(I)$, $x \mapsto \text{ad } x \mid_I$, be a representation of J, and let $C = \text{Ker } f$. Then $J/C \in \mathfrak{F}$ since $I \in \mathfrak{F}$. Clearly $C \lhd J + I$ and $J + I/C \in \mathfrak{F}$. Since $\mathfrak{F} \leq \mathfrak{L}$, we have $J/C = \langle H + C/C, K + C/C \rangle$ si $J + I/C$. Thus J si $J + I$ si L.

Corollary 1. *Let $\mathfrak{X}$ and $\mathfrak{Y}$ be $\{I,Q\}$-closed classes of Lie algebras. If $\mathfrak{X} \cap E\mathfrak{A}$ and $\mathfrak{Y} \cap E\mathfrak{A}$ are contained in $\mathfrak{F}$, then $\mathfrak{X} \mathfrak{L} \mathfrak{Y} \leq \mathfrak{L}$.*

Proof. By Theorems 1 and 2 we have $\mathfrak{L} \leq \mathfrak{X}(\mathfrak{L} \mathfrak{Y}) \leq (\mathfrak{X} \mathfrak{L})\mathfrak{Y} \leq \mathfrak{L} \mathfrak{Y} \leq \mathfrak{L}$.

The following is a generalization of [3], Corollary 5.

Corollary 2. *Let $\mathfrak{M} = \text{Max-si} \cup \text{Min-si} \cup \mathfrak{G}^I$. Then $(E\mathfrak{M}) L (E\mathfrak{M}) = \mathfrak{L}$.*

Proof. $E\mathfrak{M}$ is clearly $\{I,Q\}$-closed, and $E\mathfrak{M} \cap E\mathfrak{A} \leq \mathfrak{F}$. Hence the result follows from Corollary 1.

References

[1] R.K. Amayo and I. Stewart, *Infinite-dimentional Lie Algebras*, Noordhoff, Leyden, 1974.

[2] M. Honda, 'Joins of weak subideals of Lie algebras', *Hiroshima Math. J.* 12(1982), 657-673.

[3] M. Honda, 'Lie algebras in which the join of any set of subideals is a subideal', *Hiroshima Math. J.* 13(1983), 349-355.

[4] J.C. Lennox and S.E. Stonehewer, *Subnormal Subgroups of Groups*, Clarendon Press, Oxford, 1987.

[5] D.S. Robinson, 'Joins of subnormal subgroups', *Illinois J. Math.* 9(1965), 144-168.

[6] D.S. Robinson, 'On the theory of subnormal subgroups', *Math. Z.* 89(1965), 30-51.

[7] H. Smith, 'Groups with the subnormal join property', *Canad. J. Math.* 37(1985), 1-16.

Department of Mathematics
Faculty of Science
Hiroshima University
Hiroshima 730
Japan

Tsuwano High School
Tsuwano-cho
Shimane-ken 699-56
Japan

A NOTE ON MAXIMAL SUBGROUPS IN FINITE GROUPS

Reyadh R. Khazal

It is well known that if a finite group G has exactly one maximal subgroup then $|G|$ is divisible by one prime only and G is cyclic. In this connection one might ask whether if G has exactly two or three maximal subgroups the above result could be extended. If G has exactly three maximal subgroups then neither G needs to be cyclic nor it is required for $|G|$ to be divisible by three primes. Klein 4-group V is an example. However, in the other case, it is shown here that $|G|$ is indeed divisible by two primes only and G is cyclic. Using this fact, it is proved further that if a group G has exactly two ith maximal subgroups then all the Sylow subgroups of G are cyclic and therefore G is supersolvable. One may recall that X_i is an ith maximal subgroup of a group G if there exists a series $X_0 = G \supset X_1 \supset X_2 \supset \cdots \supset X_i$ of subgroups where X_k is a maximal subgroup of X_{k-1}, $1 \leq k \leq i$. All groups considered in this note are finite.

Lemma 1. *If a group G has exactly two maximal subgroups then G is nilpotent.*

Proof. Let M and M^* be the two maximal subgroups of G. If $M \ntrianglelefteq G$ then $N_G(M) = M$ since M is maximal and $[G : N_G(M)] \geq 2$. However $[G : N_G(M)]$ is also the number of conjugates of M in G and since a conjugate of a maximal subgroup is again a maximal subgroup it follows that $[G : N_G(M)] = 2$, i.e. $[G : M] = 2$. This implies $M \trianglelefteq G$. Similarly $M^* \trianglelefteq G$. Since all the maximal subgroups of G are normal it follows that G is nilpotent.

Lemma 2. *There exists no p–group which has exactly two maximal subgroups.*

Proof. Let P be a p-group with exactly two maximal subgroups, and let Φ be its Frattini subgroup. Then P/Φ has exactly two maximal subgroups, which are the images of the maximal subgroups of P. But P/Φ is an elementary abelian p-group, of order p^n say. Since P/Φ has two maximal subgroups, $n > 1$ and so P/Φ has an image $P/K \cong C_p \times C_p$. Thus P/K has $p + 1$ maximal subgroups, and the preimage of each of these in P is a maximal subgroup of P. This gives a contradiction.

We omit the proof of the following well known result.

Lemma 3. *If a group G has exactly one maximal subgroup then G is a cyclic p–group.*

Theorem 1. *Let G be a group which has only two maximal subgroups. Then G is cyclic and $|G|$ is divisible by two distinct primes.*

Proof. By Lemma 1, G is nilpotent and $G = P_1 \times P_2 \times \cdots \times P_m$, where P_i is the Sylow p_i-subgroup of G. We claim that $m = 2$. Suppose $m > 2$ and consider P_i (note that $m \neq 1$ by Lemma 2). If P_i does not have a proper subgroup then P_i is cyclic of a prime order and if P_i has a proper subgroup we may conclude that P_i has some maximal subgroup L_i. In either of the cases, G has a maximal subgroup $H = P_1 \times \cdots \times P_{i-1} \times P_{i+1} \times \cdots \times P_m$ in the first case and $H = P_1 \times \cdots \times P_{i-1} \times L_i \times P_{i+1} \times \cdots \times P_m$ in the second case. Thus for each i there is a maximal subgroup of G and therefore $m = 2$ and $G = P_1 \times P_2$. Evidently, neither P_1 nor P_2 can have more than one maximal subgroup. Consequently it implies that P_1 and P_2 are both cyclic and therefore G is cyclic and the theorem is proved.

The following well known theorem due to B. Huppert is used in the proof of Theorem 3. We mention it here for the sake of completeness.

Theorem 2. *If every maximal subgroup of a group G is supersolvable then G is solvable.*

Lemma 4. *If every Sylow subgroup of a group G is cyclic then G is supersolvable.*

Proof. By induction every maximal subgroup of G is supersolvable and hence G is solvable. Let N be a minimal normal subgroup of G. N is cyclic and therefore has prime order. Consider G/N. It is supersolvable by induction and so G is supersolvable.

Theorem 3. *If a group G has only two ith maximal subgroups for some integer i ($i < l = $ length of a maximal chain from e to G) then the Sylow subgroups of G are cyclic and G is supersolvable.*

Proof. Let M_k denote a kth maximal subgroup of G and consider an $(i-1)$th maximal subgroup M_{i-1}. Either M_{i-1} has no proper subgroup in which case M_{i-1} is cyclic of prime order or else M_{i-1} has at most two maximal subgroups. In either of the cases M_{i-1} is cyclic. If M_{i-2} is an $(i-2)$th maximal subgroup then it now follows that each one of its Sylow subgroup must be cyclic and therefore M_{i-2} is supersolvable. Let M_{i-j} be an $(i-j)$th maximal subgroup and assume that all the Sylow subgroups of M_{i-j} are cyclic. Every Sylow subgroup of an $(i-j-1)$th maximal subgroup M_{i-j-1} is contained in some $(i-j)$th maximal subgroup and so is cyclic. It now follows by induction that every Sylow subgroup of G is cyclic and, by Lemma 4, is supersolvable so the proof is complete.

It was remarked earlier that if a group G possesses exactly three maximal subgroups then the order of G need not be divisible by three primes. G could be a p-group. In fact for such a p-group the prime p must be 2.

Proposition. *There is no p-group for odd p with exactly three maximal subgroups.*

Proof. Let P_1, P_2, P_3 be the three maximal subgroups of a p-group P, $p \neq 2$ and $|P| = p^n$. If $P_1 \cap P_2 = \langle e \rangle$ then $|P_1 P_2| = |P|$ implies $p^n = p^2$ so that $(p^2 - 1)/(p - 1) = 3$ and $p = 2$, a contradiction. Suppose $N = P_1 \cap P_2$ and consider P/N. If $N \nsubseteq P_3$ then P/N is a p-group with exactly two maximal subgroups P_1/N and P_2/N which is impossible. Thus $N = P_1 \cap P_2 \cap P_3$ and

$|P/N| = p^{n'}$ for some integer n'. Since two maximal subgroups P_1/N and P_2/N of P/N intersect trivially, $p^{n'} = p^2$ and once again we get $p = 2$, a contradiction. Therefore the assumption that the proposition is false is wrong and the proof is complete.

The non-abelian 2-groups with some extra conditions on maximal subgroups are all classified ([1], Theorem 14.9, p.91). Evidently, abelian groups having exactly three maximal subgroups can have the order divisible by at most three primes. In fact, a group with three maximal subgroups which is not a p-group must be cyclic and its order is divisible by three primes.

Theorem 4. *A group G which has exactly three maximal subgroups and is not a group of prime power order is necessarily cyclic and its order is divisible by at most three primes.*

Proof. Let M_1, M_2, M_3 be the maximal subgroups of G. If none of $M_i \trianglelefteq G$, $i = 1, 2, 3$, then $N_G(M_1) = M_1$ and $[G : N_G(M_1)] =$ the number of conjugates of $M_1 = 3$. For if $[G : N_G(M_1)] = 2$ then $[G : M_1] = 2$ and so $M_1 \trianglelefteq G$. This however implies the indices of all the maximal subgroups are same. If $p \mid [G : N_G(M_1)] = [G : M_1]$ then there is a maximal subgroup containing a Sylow p-subgroup and its index is prime to p. Consequently, the index of M_1 is not divisible by p and we have a contradiction. Hence $M_1 \trianglelefteq G$. This implies $M_2 \trianglelefteq G$ as otherwise $[G : N_G(M_2)] = [G : M_2] =$ the number of conjugates of $M_2 = 2$ unless $M_3 \trianglelefteq G$. If $M_3 \trianglelefteq G$ then M_2 is not a conjugate of M_3 or M_1 and therefore $M_2 \trianglelefteq G$. If on the other hand $[G : M_2] = 2$, $M_2 \trianglelefteq G$ and again M_3 then is necessarily normal in G. Thus all the Sylow subgroups of G are normal and $G = P_1 \times P_2 \times \cdots \times P_m$ where P_i is a Sylow p_i-subgroup of G. Note $m \leq 3$, as otherwise G will have more than three maximal subgroups. Thus $|G|$ is divisible by at most three primes and $m = 2$ or 3. However if $m = 2$, i.e. $G = P_1 \times P_2$ then G will have either less than three or more than three maximal subgroups. This follows easily from the consideration of maximal subgroups of P_1 and P_2. Thus $m = 3$ and P_i has at most one maximal subgroup, $i = 1, 2, 3$. This however implies that G is cyclic and the proof is complete.

Corollary. *A group G which has exactly three second maximal subgroups is solvable.*

Proof. If M is any maximal subgroup of G then M has at most three maximal subgroups and therefore M is either cyclic or a 2-group. Thus every maximal subgroup of G is supersolvable and consequently G is solvable.

Remark. A group satisfying the condition in the corollary above will have cyclic Sylow subgroups corresponding to odd prime divisors of the group order. A_4 is an example of such a group.

Theorem 5. *If every minimal subgroup of a group G is complemented in G then G is supersolvable.*

Proof. Let H be any subgroup of G and $\langle a \rangle$ be a minimal subgroup in H. Then $G = \langle a \rangle T$, $\langle a \rangle \cap T = \langle e \rangle$. Consequently, $H = \langle a \rangle (H \cap T)$, $\langle a \rangle \cap (H \cap T) = \langle e \rangle$. By induction it now follows that every maximal subgroup of G is supersolvable and therefore G is solvable ([4], Theorem 2.3, p.10). Let N be a minimal normal subgroup of G. If $b \in N$ then $G = \langle b \rangle K$, $\langle b \rangle \cap K = \langle e \rangle$ and K is a maximal subgroup of G. This implies $N = \langle b \rangle (N \cap K)$ by Dedekind's modular law and N being minimal normal in G it follows that $N \cap K = \langle e \rangle$. Hence $N = \langle b \rangle$ and G/N are supersolvable which however implies that G is supersolvable.

Corollary. *If every subgroup of composite index is supplemented by a subgroup of prime index in a group G then G is supersolvable.*

Remark. The Sylow subgroups of such a group G as stated in the theorem are not necessarily cyclic as the instance of S_3 might suggest. Let

$$G = \langle a, b, x \mid a^3 = b^3 = 1, \ ab = ba, \ a^x = a^2, \ b^x = b^2, \ x^2 = e \rangle.$$

G is a group of order 18 which is supersolvable and every minimal subgroup is complemented in G. However, the Sylow 3-subgroup of G is not cyclic.

References

[1] Huppert, B., 'Normal teiler and maximale Untergruppen endlicher Gruppen', *Math. Z.*, 60(1954), 409-434.

[2] Huppert, B., *Endlichen Gruppen I*, Springer-Verlag, New York, 1967.

[3] Janko, A., 'Finite groups with invariant fourth maximal subgroups', *Math. Z.*, 82(1963), 82-89.

[4] Weinstein, M. *Between nilpotent and solvable*, Polygonal Publishing House, N.J., 1982.

Department of Mathematics
Kuwait University
P.O. Box 5969
13060 Safat
Kuwait

ON GROUPS WITH A PERMUTATIONAL PROPERTY ON
COMMUTATORS[*]

Patrizia Longobardi

1. Introduction. Let G be a group and n be an integer ≥ 2. We will say that $G \in C_n$ if and only if, for any $(x_1, x_2, \cdots, x_n) \in G^n$, there exists a non-trivial permutation σ on $\{1, 2, \cdots, n\}$ such that

$$[x_1, x_2, \cdots, x_n] = [x_{\sigma(1)}, x_{\sigma(2)}, \cdots, x_{\sigma(n)}].$$

Write $C = \cup\{C_n \mid n \geq 2\}$.

Groups G in C_2 have been studied by I.D. MacDonald in [5]. He proved that if G is in C_2 then $[\gamma_3(G), \gamma_2(G)] = 1$, $G'' \leq \zeta(G)$ and $\exp G' \leq 4$. N.Gupta and F. Levin in [1] studied the variety $V(n, \sigma, d)$ of all groups satisfying the identity

$$[x_1, x_2, \cdots, x_n] = [x_{\sigma(1)}, x_{\sigma(2)}, \cdots, x_{\sigma(n)}]^d,$$

where d is an integer and σ a *fixed* non-trivial permutation of $\{1, 2, \cdots, n\}$. Extending some previous results of H. Meier-Wunderly, E.B. Kikodze and I.D. MacDonald, they proved that $V(n, \sigma, d)$ is nilpotent-by-nilpotent if $\sigma \neq (12)$, abelian-by-nilpotent if $n > 2$, $\sigma(n) \neq n$, and nilpotent of class at most $n + 1$ if $\{1, 2\} \neq \{\sigma(1), \sigma(2)\}$.

In this paper, after some general remarks on groups in C (see §2), we study groups in C_3. We prove that a finite group of odd order in C_3 is nilpotent of class ≤ 3, actually of class ≤ 2 if 3 does not divide $|G|$ (see §3). Moreover any finite group in C_3 is p-nilpotent, for any prime $p \neq 2, 3$ (see §4).

2. General remarks. In this paragraph we give some sufficient conditions for a group to be in the class C.

2.1. *Let G be a group. If there exists $m \geq 1$ such that $G/\zeta_m(G) \in C$, then $G \in C$.*

Proof. Assume, for instance, $G/\zeta_m(G) \in C_s$. We will show that $G \in C_{s+m}$. If $(x_1, \cdots, x_s, \cdots, x_{s+m})$ is a $(s + m)$–tuple of elements of G, then there exists a non-trivial permutation σ on $\{1, 2, \cdots, s\}$ such that

$$[x_1, x_2, \cdots, x_s]\zeta_m(G) = [x_{\sigma(1)}, x_{\sigma(2)}, \cdots, x_{\sigma(s)}]\zeta_m(G).$$

---

[*] Work supported by M.P.I.

Therefore there exists $c \in \zeta_m(G)$ such that $a = cb$, where $a = [x_1, x_2, \cdots, x_s]$ and $b = [x_{\sigma(1)}, x_{\sigma(2)}, \cdots, x_{\sigma(s)}]$. Thus we get

$$
\begin{aligned}
[x_1, \cdots, x_s, \cdots, x_{s+m}] &= [a, x_{s+1}, \cdots, x_{s+m}] \\
&= [cb, x_{s+1}, \cdots, x_{s+m}] \\
&= [b, x_{s+1}, \cdots, x_{s+m}] \\
&= [x_{\sigma(1)}, \cdots, x_{\sigma(s)}, x_{s+1}, \cdots, x_{s+m}],
\end{aligned}
$$

as required. $\qquad\square$

2.2. Let G be a group. If either $G/C_G(G')$ is nilpotent or there exists a normal subgroup A of G such that G/A is nilpotent and $G/C_G(A)$ is finite, then $G \in C$.

Proof. If $G/C_G(G')$ is nilpotent, then the result follows from Lemma 3.6 of [7]. Now suppose G with a normal subgroup A such that G/A is nilpotent of class m and $G/C_G(A)$ is of finite order n. We will prove that $G \in C_{m+n+2}$. Let $(x_1, x_2, \cdots, x_m, x_{m+1}, \cdots, x_{m+n+2})$ be a $(m+n+2)$-tuple of elements of G. Obviously $[x_1, x_2, \cdots, x_{m+1}] \in A$ and there exist $i, j \in \{m+2, \cdots, m+n+2\}$ such that $i < j$ and $x_j = x_i c$, with $c \in C_G(A)$. Put $a = [x_1, x_2, \cdots, x_m, x_{m+1}]$; we get

$$
\begin{aligned}
[x_1, x_2, \cdots, x_{m+1}, x_{m+2}, \cdots, x_j] &= [a, \cdots, x_i, \cdots, x_j] \\
&= [a, \cdots, x_i, \cdots, c][a, \cdots, x_i, \cdots, x_i]^c \\
&= [a, \cdots, x_i, \cdots, x_i].
\end{aligned}
$$

and

$$
\begin{aligned}
[x_1, x_2, \cdots, x_{m+1}, \cdots, x_j, \cdots, x_i] &= [a, \cdots, x_j, \cdots, x_i] \\
&= [a, \cdots, x_i c, \cdots, x_i] \\
&= [[a, \cdots, c][a, \cdots, x_i]^c, \cdots, x_i] \\
&= [a, \cdots, x_i, \cdots, x_i].
\end{aligned}
$$

Hence $[x_1, \cdots, x_{m+1}, \cdots, x_i, \cdots, x_j] = [x_1, \cdots, x_{m+1}, \cdots, x_j, \cdots, x_i]$ and G belongs to C_{m+n+2}. $\qquad\square$

3. Finite groups of odd order in C_3.

We will prove that a finite group G of odd order in C_3 is nilpotent of class ≤ 3, actually of class ≤ 2 if 3 does not divide $|G|$.

Lemma 3.1. Let $G \in C_3$ be a finite metabelian group such that $[G', G]$ is of odd order, and let $a, b, c \in G$.

i) If $[a, b, c]$, $[b, c, a]$, $[c, a, b] \neq 1$, then $[a, b, c] = [b, c, a] = [c, a, b]$.

ii) If $[a, b, c] \neq 1$ and $[b, c, a] = 1$ (respectively $[c, a, b] = 1$), then $[a, b, c] = [a, c, b]$ (respectively $[a, b, c] = [c, b, a]$).

Proof. It immediately follows from the equality: $[a, b, c][b, c, a][c, a, b] = 1$ and from $[c, a, b]^{-1} = [a, c, b]$. $\qquad\square$

Lemma 3.2. *Let $G \in C_3$ be a finite group with $[G', G]$ of odd order, $a_1, a_2, a_3 \in G$ and N be a normal subgroup of G with G/N metabelian.*

i) *If $[a_1, a_2, a_3] \notin N$, then $[a_1, a_2, a_3]N = [a_{\sigma(1)}, a_{\sigma(2)}, a_{\sigma(3)}]N$ implies*
$$[a_1, a_2, a_3] = [a_{\sigma(1)}, a_{\sigma(2)}, a_{\sigma(3)}].$$

ii) *If $[a_1, a_2, a_3] \in N$, then either $[a_1, a_2, a_3] = 1$ or $[a_2, a_3, a_1]$, $[a_3, a_1, a_2] \in N$.*

Proof. i) Assume $[a_1, a_2, a_3] \notin N$ and $[a_1, a_2, a_3]N = [a_{\sigma(1)}, a_{\sigma(2)}, a_{\sigma(3)}]N$. If $[a_2, a_3, a_1] \in N$, we have $[a_1, a_2, a_3]N = [a_1, a_3, a_2]N$ and $[a_1, a_2, a_3]N \neq [a_{\tau(1)}, a_{\tau(2)}, a_{\tau(3)}]N$, for any non-trivial permutation τ on $X = \{1, 2, 3\}$ different from (23), since $[G', G]$ is of odd order. Therefore, since $G \in C_3$, we get $[a_1, a_2, a_3] = [a_1, a_3, a_2]$, as required. Analogously if $[a_3, a_1, a_2] \in N$. Suppose now $[a_{\tau(1)}, a_{\tau(2)}, a_{\tau(3)}] \notin N$, for any permutation τ on X. By Lemma 3.1 i), we have $[a_1, a_2, a_3]N = [a_2, a_3, a_1]N = [a_3, a_1, a_2]N$ and $[a_1, a_2, a_3] = [a_2, a_3, a_1] = [a_3, a_1, a_2]$. Since $[G', G]$ is of odd order we get $[a_1, a_2, a_3]N \neq [a_{\eta(1)}, a_{\eta(2)}, a_{\eta(3)}]N$ for any non-trivial permutation η on X different from (123) and (132), from which the conclusion.

ii) It easily follows from i) and from $[G', G]$ of odd order. $\square$

Lemma 3.3. *Let $G = A\langle x \rangle \in C_3$, where $A \triangleleft G$ abelian and x of odd order. Then $[a, x, x, x] = 1$ and $[a, x, x]^3 = 1$, for every $a \in A$.*

Proof. G is metabelian, so for every $a \in A$ we have $[x, ax, ax^2][ax, ax^2, x][ax^2, x, ax] = 1$. Moreover $[x, ax, ax^2] = [x, a, x^2]^x$, $[ax, ax^2, x] = [[a, x^2]^x[x, a]^x, x] = [a, x^2, x]^x[x, a, x]^{x^2}$ and $[ax^2, x, ax] = [a, x, x]^{x^2}$. If either $[a, x, x] = 1$ or $[x, a, x^2] = 1$, then the conclusion is obvious. Now assume $[ax, ax^2, x] = 1$, so that $[a, x^2, x]^x = [a, x, x]^{x^2}$. Then $[a, x]^x[x^2, a] = [x, a]$ centralizes x, from which the result follows. Hence we can suppose $[x, ax, ax^2]$, $[ax, ax^2, x]$ and $[ax^2, x, ax]$ different from 1. By Lemma 3.1 i) we have $[x, a, x^2]^x = [a, x^2, x]^x[x, a, x]^{x^2} = [a, x, x]^{x^2}$ and so $([x, a, x^2]^x)^2 = [x, a, x^2]^x[a, x^2, x]^x[x, a, x]^{x^2} = [x, a, x]^{x^2} = [a, x, x^2]^x$, since $[x, a, x^2][a, x^2, x] = 1$. Therefore $[x, a, x^2]^3 = 1 = [a, x, x]^3$. Moreover $[a, x, x][a, x, x]^x = [[a, x][a, x]^x, x] = [a, x^2, x] = [a, x, x^2] = [x, a, x^2]^{-1} = [x, a, x]^x$, from which $[a, x, x] = ([x, a, x]^2)^x = ([x, a, x]^{-1})^x = [a, x, x]^x$, so that $[a, x, x, x] = 1$, as required. $\square$

Lemma 3.4. *Let $G = \langle a, b \rangle$ be a finite metabelian group of odd order. If G is in C_3, then $|[G', G]| \leq 3$.*

Proof. First assume $[a, b, b] = 1$; hence $\langle b \rangle^G$ is abelian[1] and Lemma 3.3 applied to $G = \langle b \rangle^G \langle a \rangle$ leads to $[b, a, a]^3 = 1$ and $[b, a, a, a] = 1$. We also have $[b, a, a, b] = [b, a, b, a] = 1$, from which $[G', G] = \langle [a, b, a] \rangle$ is of order ≤ 3, as required. The result analogously follows if $[b, a, a] = 1$ or $[a, b, ab] = 1$ or $[a, b, ba] = 1$. Now suppose $[a, b, b]$, $[a, b, a]$, $[a, b, ab]$, $[a, b, ba]$ different from 1. Then we have $[a, b, ab] = [a, b, b][a, b, a]^b \neq 1$, $[b, ab, a] = [[b, a]^b, a] = [b, a, a][b, a, b, a] =$[2] $[b, a, a][b, a, a, b] =$

---

[1] $[b, a, b] = 1$ implies $[b, b^a] = 1$ and $1 = [b^a, b, a][b, a, b^a][a, b^a, b] = [a, b^a, b] = [b^a, a, b]$. We also have $[b, a^2, b] = [b, a, b]^{[b,a]^a}[b^a, a, b] = 1$, so that $[b, a^{a^2}] = 1$. Assume now $[b, b^{a^i}] = 1$. We get $[b, a^{i+1}, b^a] = [b, a, b^a]^{[b,a^i]^a}[b, a^i, b]^a = 1$ and $1 = [b, a^{i+1}, b^a][a^{i+1}, b^a, b][b^a, b, a^{i+1}] = [a^{i+1}, b^a, b]$, so b permutes with $b^{a^{i+2}}$. Then $[b, b^{a^n}] = 1$ for any $n \in \mathbf{N}$.

[2] G is metabelian, so $[x, y, z, t] = [x, y, t, z]$ for any x, y, z, t in G.

$[b, a, a]^b \neq 1$ and $[ab, a, b] = [b, a, b] \neq 1$ and Lemma 3.1 i) implies $[a, b, b][a, b, a]^b = [b, a, a][b, a, a, b] = [b, a, b]$.

Arguing analogously with $[a, b, ba]$, $[b, ba, a]$ and $[ba, a, b]$, we also get

$$[a, b, a][a, b, b]^a = [b, a, b][b, a, b, a] = [b, a, a].$$

Then $[b, a, b] = [b, a, a][b, a, a, b] = [b, a, b][b, a, a, b]^2$, from which $[b, a, a, b] = 1$ and $[b, a, b] = [b, a, a]$. Finally $[b, a, b]^3 = [a, b, b][a, b, a][b, a, b][b, a, a] = 1$, $[b, a, b] \in \zeta(G)$ and $[G', G] = \langle [b, a, b] \rangle$ is of order ≤ 3, as required. $\square$

Theorem 3.5. *Let G be a finite group of odd order such that 3 does not divide $|[G', G]|$. Then $G \in C_3$ if and only if $[G', G] = 1$.*

Proof. First assume G metabelian and consider $x, y \in G$. Put $H = \langle x, y \rangle$, we have $|[H', H]| \leq 3$ by Lemma 3.4, and so $[H', H] = 1$ and $[x, y, y] = 1$. Therefore G is a 2-Engel group and by a result of F.W. Levi (see [4]), $[G', G]^3 = 1$; hence the hypothesis implies $[G', G] = 1$. Now we argue by induction on the derived length of G. Let $a, b \in G$ and $K = \langle a, b \rangle$. K/K'' is metabelian, so $K'/K'' \leq \zeta(K/K'')$ and $K'/K'' = \langle [a, b]K'' \rangle$ is cyclic. By induction we get $K'' \leq \zeta(K')$ and so K is metabelian. Hence $[K', K] = 1$ and $[a, b, b] = 1$. The group G is 2-Engel and again $[G', G] = 1$. The converse is obvious. $\square$

Remark 3.6. *Let $G = A \rtimes B$, where A is the unique minimal normal subgroup of G, $|A| = 3^n$ $(n > 1)$, A is elementary abelian and B is of prime order $q > 3$. Then $G \notin C_3$.*

Proof. G is a metabelian group of odd order, 2-generator with $[G', G] = A$. Then the conclusion follows from Lemma 3.4. $\square$

Theorem 3.7. *Let G be a finite group of odd order. Then, if G is in C_3, G is nilpotent.*

Proof. By Theorem 3.5 we can assume that 3 divides $|[G', G]|$. By contradiction suppose the result false, and let G be a minimal counterexample. Thus G has a unique minimal normal subgroup N, G/N is nilpotent and N is the last non-trivial term of the lower central series of G. N is elementary abelian, so $N \cap \Phi(G) = 1$ implies $G = N \rtimes H$ for a suitable $H \leq G$, by a well-known result of W. Gaschütz. Therefore H is nilpotent and N is a p-group. p does not divide $|H|$, otherwise $H = H_p \times H_{p'}$, with $H_p \neq 1$, and so $|N H_{p'}| < |G|$, $N H_{p'}$ is nilpotent and $N = [N, H] = [N, H_p] < N \leq N H_p$. If $H = H_q \times H_{q'}$ with $H_q \neq 1$ and $H_{q'} \neq 1$ (q a prime), then as before $N H_q$ and $N H_{q'}$ are nilpotent, hence $[N, H] = 1$, a contradiction. Therefore, H is a q-group (q a prime) and, by a similar argument, it has a unique maximal subgroup. Thus it is easy to see that H is cyclic, of prime order. Hence $G = N \rtimes H$, $N = G'$, $|N| = p^n$ for a suitable integer n. Therefore $p = 3$, $n > 1$ and $G \notin C_3$ by Remark 3.6. $\square$

Proposition 3.8. *Let G be a metabelian group. If $[G', G]$ is of order 3, then G is in C_3.*

Proof. Let $x, y, z \in G$ and $[G', G] = \{1, a, b\}$. If $[x, y, z] = 1$, then $[y, x, z] = 1$ and so $[x, y, z] = [y, x, z]$. Now assume $[x, y, z] \neq 1$ and put $a = [x, y, z]$. Hence $[y, x, z] = [x, y, z]^{-1} = a^{-1} = b$.

If $[y, z, x] \neq 1$, then $\{[y, z, x], [z, y, x]\} = \{a, b\}$; if $[y, z, x] = 1$, then $1 = [x, y, z][y, z, x][z, x, y]$ $= [x, y, z][z, x, y]$ and $[x, y, z] = [x, z, y]$. Therefore G is in C_3.   □

Theorem 3.9. *Let* $G = \langle a, b \rangle$ *be a finite 3-group. Then* G *is in* C_3 *if and only if* $\|[G', G]\| \leq 3$.

Proof. G/G'' is metabelian, so $\|[G', G]/G''\| \leq 3$ by Lemma 3.4. Hence from $G'/G'' = \langle [a, b]G''$, $[G', G]/G'' \rangle$ it follows G'/G'' is 2-generator and so G' 2-generator. Therefore G is metabelian (see [2], Theorem III 7.9) and again Lemma 3.4 implies $\|[G', G]\| \leq 3$.

Conversely, by $\|[G', G]\| \leq 3$, G is nilpotent of class ≤ 3 and so metabelian. Then the conclusion follows from Proposition 3.8.   □

Lemma 3.10. *Let* $G = \langle x, y, z \rangle$ *be a finite 3-group of class* ≤ 3. *If* G *is in* C_3, *then* $\|[G', G]\| \leq 3$.

Proof. G is a metabelian group with $[G', G] \leq \zeta(G)$. If G is a 2-Engel group, then $[x, y, z] = [x, z, y]^{-1}$ and $[G', G]^3 = 1$, which mean $[G', G] = \langle [x, y, z] \rangle$ of order ≤ 3, as required.

Assume now that G is not a 2-Engel group. Then we can suppose $[x, y, z] \neq 1$.[3] We can also assume that $[x, y, x] \neq 1$[4] and $[x, y, z]$, $[y, z, x]$, $[z, x, y]$ different from 1.[5] Then by Lemma 3.1 we get $[x, y, z] = [y, z, x] = [z, x, y] = 1$ and $[x, y, z]^3 = 1$. Now we will prove that $[G', G] = \langle [x, y, z] \rangle$, which implies the required result.

Consider $[x, y, yz] = [x, y, z][x, y, y]$. If $[x, y, yz] = 1$, then $[x, y, y] \in \langle [x, y, z] \rangle$. Now suppose $[x, y, yz] \neq 1$, so $[x, y, yz] \neq [x, y, yz]^{-1} = [y, x, yz]$. We also have $[x, y, yz] \neq [x, yz, y] = [x, y, y][x, z, y]$ otherwise $[x, y, z] = [x, z, y] = [z, x, y]^{-1} = [x, y, z]^{-1}$. Moreover $[x, y, yz] \neq [y, yz, x] = [y, z, x] = [x, y, z]$, since $[x, y, y] \neq 1$, and $[x, y, yz] \neq [yz, x, y] = [y, x, y][z, x, y]$, otherwise $[x, y, y] = [y, x, y] = [x, y, y]^{-1}$. Then, from $G \in C_3$, it follows that $[x, y, yz] = [yz, y, x] = [z, y, x]$, so that $[x, y, y] = [y, z, x]^{-2} \in \langle [x, y, z] \rangle$.

Analogously, arguing on $[y, zx, x]$, $[x, yz, z]$, $[z, x, xy]$, $[y, z, xz]$ and $[z, xy, y]$, we get $[y, x, x]$, $[x, z, z]$, $[z, x, x]$, $[y, z, z]$, $[z, y, y] \in \langle [x, y, z] \rangle$.   □

Theorem 3.11. *Let* $G = \langle x, y, z \rangle$ *be a finite 3-group. Then* $G \in C_3$ *if and only if* $\|[G', G]\| \leq 3$.

Proof. Assume $G \in C_3$. We will prove by induction on $|G|$ that G is of class ≤ 3; so the required

---

[3] G is not a 2-Engel group, so at least one of the generators, say x, is not a right Engel element. If $[x, y, y] = 1 = [x, z, z] = [x, xy, xy] = [x, xz, xz]$, then $[x, y, x] = 1 = [x, z, x]$. Therefore we have $[x, yz, yz] \neq 1$, otherwise $[x, y, z] = [x, z, y]^{-1}$, $[x, x^n y^m z^t, x^n y^m z^t] = [x, y, z]^{mt}[x, z, y]^{mt} = 1$ for every integers n, m, t and so $[x, a, a] = 1$ for every $a \in G$, a contradiction.

[4] If $[x, y, x] = 1$, we have $[yx, y, yx] = [x, y, y][x, y, x] = [x, y, y] \neq 1$ and $[yx, y, y] = [x, y, y] \neq 1$, so we can replace x by yx.

[5] If $[x, y, z] = 1$, then $[x, y, xz] = [x, y, z][x, y, x] = [x, y, x] \neq 1$, so we can replace z by xz. Suppose now $[y, z, x] = 1$. We have $[y, zx, x] = [y, x, x] \neq 1$ and $[y, zx^2, x] = [y, x, x]^2 \neq 1$. Moreover $[x, y, zx] = [x, y, x][x, y, z]$ and $[x, y, zx^2] = [x, y, x]^2[x, y, z]$, so at least one of the two commutators $[x, y, zx]$ and $[x, y, zx^2]$ is non-trivial. Hence we can suitably replace z either by zx or by zx^2.

Finally suppose $[z, x, y] = 1$. We ger $[zy, x, y] = [y, x, y] \neq 1$ and $[zy^2, x, y] = [y, x, y]^2 \neq 1$.

Moreover $[x, y, zy] = [x, y, y][x, y, z]$ and $[x, y, zy^2] = [x, y, y]^2[x, y, z]$, so at least one of these is different from 1. Finally $[y, zy, x] = [y, z, x] \neq 1$ and $[y, zy^2, x] = [y, z, x] \neq 1$, so we can replace z either by zy or by zy^2.

assertion will follow from Lemma 3.10. Let N be a normal subgroup of G, with $N \leq G'$ and $|N| = 3$. Then we have $G/N = \langle xN, yN, zN \rangle$, and $|[G', G]/N| \leq 3$; we can assume[6] $[x, y, z] \notin N$, which leads to $[G', G] = \langle [x, y, z], N \rangle$. Now we will prove that $[x, y, z, x] = 1$. By Theorem 3.9 we have $[a, x, x, x] = 1$ for every $a \in G$. If there exists $b \in G$ such that $[b, x, x] \notin N$, then $\langle [b, x, x] N \rangle = [G', G]/N = \langle [x, y, z] N \rangle$ from which $[x, y, z, x] = 1$.

Therefore we can suppose $[a, x, x] \in N$ for every $a \in G$. If $[z, x, y] \in N$, we get $[x, y, z] N = [z, y, x] N$ and $[z, y, xz] N = [xz, y, x] N$, since G/N is metabelian. Thus Lemma 3.2 implies $[x, y, z] = [z, y, x]$ and $[z, y, xz] = [xz, y, x]$. But $[x, y, xz] = [x, y, z][x, y, x]$ and $[xz, y, x] = [x, y, x][x, y, z, x][z, y, x]$, since $[y, x, x] \in N \leq \zeta(G)$; so $[x, y, z, x] = 1$.

Suppose now $[z, x, y] \notin N$. Then for a suitable $\varepsilon \in \{1, -1\}$ we have $[x, y, z] N = [x, z, y]^{\varepsilon} N$, $[x, y, zz] N = [x, zx, y]^{\varepsilon} N$, $[x, y, xz] N = [x, xz, y]^{\varepsilon} N$ and so by Lemma 3.2[7] $[x, y, z] = [x, z, y]^{\varepsilon}$, $[x, y, zx] = [x, zx, y]^{\varepsilon}$ and $[x, y, xz] = [x, xz, y]^{\varepsilon}$. Therefore we obtain

$$[x, y, z] = [x, z, y]^{\varepsilon} = [x, xz, y]^{\varepsilon} = [x, y, xz] \quad \text{and} \quad [x, y, zx] = [x, zx, y]^{\varepsilon} = [x, z, y]^{\varepsilon},$$

from which $[x, y, xz] = [x, y, zx]$ and $[x, y, z]^x = [x, y, z]$, as required.

Arguing analogously we can prove that $[y, x, z, y] = 1 = [z, x, y, z] = [z, y, x, z]$. Therefore $[x, y, z] = [y, x, z]^{-1}$ permutes with y and also with z, since[8] either

$$[G', G] = N\langle [z, x, y] \rangle \quad \text{or} \quad [G', G] = N\langle [z, y, x] \rangle.$$

Thus $[G', G] \leq \zeta(G)$ and G is nilpotent of class ≤ 3. □

Corollary 3.12. *Let G be a finite group of odd order. If $G \in C_3$, then G is nilpotent of class ≤ 3.*

Proof. G is nilpotent by Theorem 3.7. We can suppose G 3-group (see Theorem 3.5). Any 3-generator subgroup of G is nilpotent of class ≤ 3 by Theorem 3.11, so G is of class ≤ 3. □

4. Finite groups in C_3.

Lemma 4.1. *Let $G = \langle a, b \mid a^b = a^{-1} \rangle$. If $G \in C_3$, then the order of a divides 48.*

Proof. We have $[b, ab, a^2 b] = [b, a, b]^b = a^4$, $[ab, a^2 b, b] = [a, b, b]^b [b, a^2, b]^b = a^{-4} a^8 = a^4$ and $[a^2 b, b, ab] = [a^2, b, b]^b = a^{-8}$. From G metabelian and $G \in C_3$ it follows that either $a^{-8} = 1$ or $a^{-8} \in \{a^4, a^{-4}, a^8\}$. Hence either $a^{16} = 1$ or $a^{12} = 1$ and so $a^{48} = 1$. □

Theorem 4.2. *Let G be a finite group. If $G \in C_3$, then G is p-nilpotent for any prime $p \neq 2, 3$.*

---

[6] Suppose G/N of class 3. We have $G'/N \not\subseteq \zeta(G/N)$ and so, for instance, $[x, y] N \notin \zeta(G/N)$. If $[x, y, z]$, $[x, y, xz]$, $[x, y, yz] \in N$, then also $[x, y, x]$ and $[x, y, y]$ belong to N, from which the contradiction $[x, y] N \in \zeta(G/N)$. Hence, replacing eventually z either by xz or by yz, we can assume $[x, y, z] \notin N$.

[7] G is nilpotent of class ≤ 4, so $[x, y, z, [x, y]] = 1$ and $[x, y, z]^{-1} = [y, x, z]$ for any $x, y, z \in G$.

[8] If $[z, x, y] \in N$, then $[x, y, z][y, z, x][z, x, y] N = N = [x, y, z][y, z, x] N$, since G/N is metabelian. Hence $[x, y, z] N = [z, y, x] N$.

Proof. Let P be a p-subgroup of G with $p \neq 2, 3$. We will prove that $N_G(P)/C_G(P)$ is a p-group. If there exists $x \in N_G(P) - C_G(P)$ such that $|x|$ is a prime different from p, then if $|x| \neq 2$, we have $p(x) = P \times \langle x \rangle$ (see Theorem 3.5 or Theorem 3.7) and the contradiction $x \in C_G(P)$. Thus suppose $|x| = 2$. If $a \in P$, we get $[a.x] \in P$, $1 = [a, x^2] = [a, x][a, x]^x$ and $[a, x]^x = [a, x]^{-1}$. Hence Lemma 4.1 yields $[a, x]^{48} = 1$ and so $[a, x] = 1$. Therefore $x \in C_G(P)$, a contradiction. Thus $N_G(P)/C_G(P)$ is a p-group and, by Frobenius' Theorem, G has a normal p-complement.

References

[1]   N. Gupta and F. Levin, 'Some symmetric varieties of groups', *Bull. Austral. Math. Soc.*, 3(1970), 97-105.

[2]   B. Huppert, *Endliche Gruppen I*, Springer-Verlag, Berlin, 1967.

[3]   E.B. Kikodze, 'Some identities in groups', *Math. USSR Izvestija*, 1(1967), 253-258.

[4]   F.W. Levi, 'Groups in which the commutator operation satisfies certain algebraic conditions', *J. Indian Math. Soc.*, 6(1942), 87-97.

[5]   I.D. MacDonald, 'On certain varieties of groups', *Math. Zeitschr.*, 76(1961), 270-282.

[6]   H. Meier-Wunderli, 'Über die Gruppen mit der identischen Relation $(x_1, \cdots, x_n) = (x_2, \cdots, x_n, x_1)$ $n \geq 3$', *Vjschr, naturf. Ges. Zürich*, 94(1949), 211-218.

[7]   S.J. Tobin, 'Groups with exponent 4', *Proceedings of Groups – St. Andrews 1981*, L.M.S. Lecture Note Series 71, Cambridge University Press, 1982.

Dipartimento di Matematica e Applicazioni
'R. Caccioppoli"
Mezzocannone, 8
80134 Napoli
Italy

ON FIBONACCI GROUPS AND SOME OTHER GROUPS

Jens L. Mennicke

0. The present article reports on some joint work with H. Helling and A.C. Kim, and with F. Grunewald and L. Vaserstein. It is an extended version of an address given to the Conference "Groups – Korea 1988". The author acknowledges financial support from Deutsche Forschungsgemeinschaft and from Korea Science and Engineering Foundation. Cordial thanks go to the organisers for excellent hospitality.

1. We report on some work on Fibonacci groups. A preliminary note [4], which was published by a misunderstanding, contains a mistake for which the present author assumes the responsibility. The note should be disregarded.

The Fibonacci groups are defined by the presentation

$$F(2,m) = \langle x_1, x_2, \cdots, x_m \; ; \; x_i x_{i+1} = x_{i+2}, \; i \bmod m \rangle.$$

These groups were introduced, in a special case, by J. Conway [2]. They were investigated by several authors, see [1] for a more complete reference. The main problem was the finiteness problem. It is now known that the groups are finite for $m = 2$, 3, 4, 5, 7, and infinite for all other values of m. The case $m = 9$, which had been open for some time, was recently treated successfully by M.F. Newman [5].

Our emphasis lies on different aspects. There is a sharp distinction between m even and m odd. Our approach does not work for m odd, and it is open, in fact unlikely, that our results hold for m odd. For the rest of this article, we shall assume that $m = 2n$ is even.

It is not difficult to see that $F(2,6)$ is a Euclidean cristallographic group, i.e. it is a finite extension of $\mathbf{Z}^3$, and a discrete subgroup of the affine orthogonal group operating on the Euclidean space $\mathbf{R}^3$. There is a similar situation for $2n > 8$, with the Euclidean space $\mathbf{R}^3$ replaced by the three-dimensional hyperbolic space.

Theorem 1: *For $n \geq 4$, there is a semiregular tessellation of hyperbolic 3-space. A fundamental domain is a polyhedron bounded by $4n$ regular triangles which are all congruent. For $n = 5$, the regular icosahedron occurs in the family. In the tessellation, for each edge there are three polyhedra which meet in this edge. The three angles are pairwise different, except for $n = 5$, where all angles are $120°$.*

For large n, the polyhedron looks like a flat disc. In fact, the diameter of the disc tends to ∞ and the height tends to 0 as $n \to \infty$.

Theorem 2: *The group $F(2,2n)$ acts as a group of isometries on the tessellation described in Theorem 1. It operates freely and sharply trasitively. The fundamental domain is a closed compact orientable 3-manifold M_n, and*

$$\pi_1(M_n) = F(2,2n).$$

The manifold M_n comes with a natural triangulation. It is known that the commutator quotient group of $F(2,2n)$ is finite:

$$F(2,2n)^{\text{ab}} = \begin{cases} \mathbf{Z}_N \times \mathbf{Z}_{5N}, & N = f_{n'} \cdot g_{n'}, & \text{for } n = 2n'; \\ \mathbf{Z}_N \times \mathbf{Z}_N, & N = g_{2n'+1}, & \text{for } n = 2n'+1. \end{cases}$$

Here f_n, g_n are the Fibonacci-Lucas numbers defined by the equation

$$h_{n+2} = h_{n+1} + h_n$$

and the initial values

$$f_0 = 0, \quad f_1 = 1 \quad \text{and} \quad g_0 = 2, \quad g_1 = 1.$$

Via Theorem 2, this means that the manifolds M_n have a finite non-trivial first integral homology group

$$H_1(M_n, \mathbf{Z}) = F(2,2n)^{\text{ab}}.$$

It is not unlikely that our approach opens an access to a study of the relationship between more subtle topological invariants such as the self-intertwining numbers in intersection theory, and objects from number theory.

Theorem 2 implies that $F(2,2n)$ has a faithful representation as a discrete subgroup $\Gamma < \mathrm{SL}_2(\mathbf{C})$ such that the orbit space

$$M_n = \Gamma \backslash \mathrm{SL}_2(\mathbf{C}) / SU_2$$

is compact, and

$$\Gamma = F(2,2n).$$

We have studied the arithmetical nature of Γ. Here is our result.

Theorem 3: *The group $\Gamma = F(2,2n)$, viewed as a discrete subgroup of $\mathrm{SL}_2(\mathbf{C})$, has integral traces, which belong to the number field*

$$T = \mathbf{Q}(\alpha, i\sqrt{\omega}),$$

where $\alpha = \cos\frac{2\pi}{n}$, and $\omega = (1+2\alpha)(3-2\alpha)$. There is a quaternion algebra Q defined over T which is a skewfield. Q has an order A such that Γ is a subgroup of the group A^* of units of A. The group

Γ *is arithmetic, in particular of finite index in A^*, for $n = 4, 5, 6, 8, 12$, and Γ is non-arithmetic for all other values of $n \geq 4$.*

It seems that the non-arithmetic groups Γ are not commensurable to Coxeter groups, except for finitely many n, the reason being that the angles occurring in the tessellation described in Theorem 1 are not rational submultiples of 2π, except for some small values of n. However, we have not studied this point in detail.

It seems that some aspects of our work broadly generalise. There is a large class of groups, defined via presentations, and containing the Fibonacci groups $F(2, 2n)$, which seem to have a faithful representation as a discrete subgroup $\Gamma < SL_2(\mathbf{C})$ with compact orbit space. We do not know, however, whether any of the geometric aspects contained in Theorem 1 carry over to this larger class of groups.

The proofs of Theorems 1, 2, 3 will appear shortly. The study of the larger class of groups is still work in progress.

2. We shall outline the proofs of Theorems 1, 2, 3.

figure 1 (next page)

The configuration displayed in figure 1 was found by the authors in special cases, and then by C.C. Simms in general.

Consider the combinatorial polyhedron shown in figure 1. Identify pairs of faces which carry the same label, obtaining a complex M_n. In general, the resulting complex will not be a manifold. In the vertices, a neighbourhood will be a star bounded by some surface of genus $n \geq 0$. There is a simple criterion due to Seifert and Threlfall. The resulting complex is a manifold if and only if its Euler characteristic vanishes. One must establish that the cirterion applies, and that M_n is a manifold.

The 2-skeleton of M_n is the union of the faces F_i. It follows that the group $F(2, 2n)$ is the fundamental group of M_n:
$$\pi_1(M_n) = F(2, 2n).$$
After we had completed our work, it was discovered independently by J. Howie and by J. Montesinos that the manifolds M_n are cyclic coverings of the 3-sphere S^3, branched over the figure-8-knot. In fact, consider the axis QR in figure 1. After the identification, this axis becomes the figure-8-knot.

The next step is to realise the combinatorial polyhedrong shown in figure 1 as a polyhedron in hyperbolic 3-space. One decomposes the polyhedron into tetrahedra. Using some hyperbolic geometry, one shows that the polyhedron can be realised as an isohedral polyhedron, bounded by $4n$ regular triangles which are all congruent under isometries of hyperbolic 3-space.

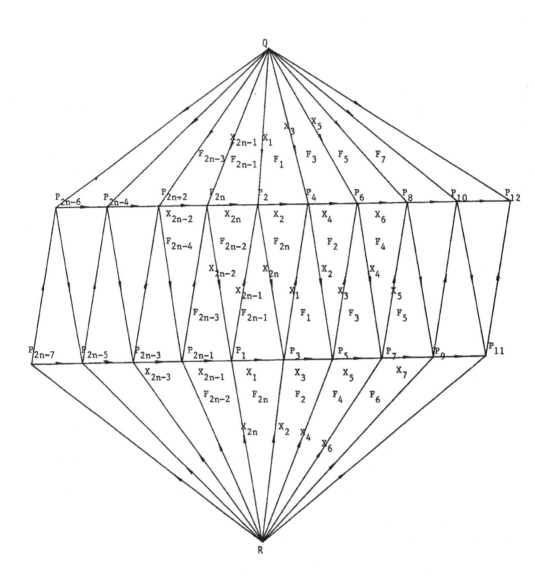

figure 1

This isohedral polyhedron is the generalisation of a very classical object: the regular icosahedron. The polyhedron depends on one real parameter.

The next step is to tessellate hyperbolic 3-space with copies of the polyhedron in such a way that the above mentioned abstract identifications of faces are isometries of hyperbolic 3-space. Using a theorem of Poincaré which was (re) proved by B. Maskit, it suffices to establish that the sum of the angles between adjacent faces, summed around each edge, is 360°. this condition produces a polynomial equation for the real parameter. This equation has precisely one solution which is geometrically meaningful, i.e. for which the angle conditions hold.

The isohedral polyhedron satisfying the angle condition can now be imbedded into hyperbolic 3-space such that all vertices have coordinates which are algebraic numbers, belonging to a numberfield arising from the nth cyclotomic number field by one or two quadratic extensions. The angle condition guarantees that hyperbolic 3-space can be tessellated with copies of the isohedral polyhedron. For each pair of faces such as shown in figure 1, there is precisely one orientation preserving isometry which maps one face onto the other.

One can view the tessellation thus obtained as a metric realisation of the universal covering of M_n. The above mentioned isometries generate a discrete subgroup of $SL_2(\mathbf{C})$ which is the monodromy group of the universal covering. The monodromy group is in turn isomorphic to the fundamental group, and hence to $F(2, 2n)$.

It is not difficult to check that the isometries generate a discrete subgroup of $SL_2(\mathbf{C})$ which has integral traces, belonging to the number field $T = Q(\alpha, i\sqrt{\omega})$, where $\alpha = \cos 2\pi/n$, and $\omega = (1 + 2\alpha)(3 - 2\alpha)$.

Next we study abstract homomorphisms of $F(2, 2n)$ into $SL_2(\mathbf{C})$. We show that such a homomorphism is uniquely determined, up to conjugacy in $SL(\mathbf{C})$, by the traces of the generators x_1, x_2, x_3. Hence we obtain an abstract characterisation of the above mentioned isomorphism between $F(2, 2n)$ and a discrete subgroup of $SL_2(\mathbf{C})$.

We use this abstract characterisation as follows. We imbed $F(2, 2n)$ into the group of units of det $= +1$ of an order of a certain quaternion algebra defined over the trace field T. The group of units of det $= +1$ acts as an arithmetic group on a (in general higher-dimensional) space defined via the various infinite valuations of T. Studying these infinite valuations of T, we find that the group $F(2, 2n)$ is commensurable to the full group of all units of det $= +1$ only for $n = 4$, 5, 6, 9, 12. For all other values of $n \geq 7$, the group $F(2, 2n)$ is non-arithmetic. It is, however, weakly arithmetic in the sense that it is a subgroup, of infinite index, in an arithmetic group. The proof uses Eichler's Strong Approximation Theorem.

It was observed by J. Howie that our work is closely related to some work of W. Thurston and of R. Riley who studied hyperbolic structures on the complement of the figure-8-knot. Thurston has defined a parametrisation for these hyperbolic structures. The parameter domain is a subset of the plane $\mathbf{R}^2$ which is the complement of a closed compact set containing 0. This set is called the structure gap. J. Howie observed that our method exhibits points which do not belong to the structure gap, in particular the point $(4,0)$ in Thurston's notation. Thurston's method of ideal tetrahedra does not produce a hyperbolic structure for this point. A modification of our argument produces hyperbolic structures for all points $(x,0)$ for $x > 3$. On the real axis, this is best possible, since the point $(3,0)$ does belong to the structure gap. Howie also observed that the structure gap is not convex, since the points $(4,1)$ and $(4,-1)$ do belong to the structure gap.

At present, it is not clear whether further modifications of our method will produce more information about the structure gap off the real axis.

3. We report on some joint work with F. Grunewald and L. Vaserstein, which belongs to unstable K-theory.

Consider the ring $\mathbf{Z}[x]$. It seems that little was known about the group

$$G = \mathrm{SL}_2(\mathbf{Z}[x]),$$

and on related groups. Here are some of our results.

Theorem 4: *The group*

$$H = \mathrm{Sp}_4(\mathbf{Z}[x])$$

is generated by elementary matrices.

This generalises some work of A. Suslin, who proved that the groups $\mathrm{SL}_n(\mathbf{Z}[x])$ are generated by elementary matrices, for $n \geq 3$.

Theorem 5: *Let k be a complex quadratic number field with maximal order of integers $o = \mathbf{Z}[\omega]$. The natural ring homomorphism*

$$\varphi : \mathbf{Z}[x] \longrightarrow \mathbf{Z}[\omega], \quad \varphi(x) = \omega$$

induces a group homomorphism

$$\varphi : \mathrm{SL}_2(\mathbf{Z}[x]) \longrightarrow \mathrm{SL}_2(\mathbf{Z}[\omega]).$$

The group homomorphism φ is a surjective homomorphism.

Notice that the surjectivity asserted in Theorem 5 for orders $\mathbf{Z}[\omega]$ other than orders in complex quadratic number fields follows from known work, in particular of L. Vaserstein.

Combined with some earlier work of R. Zimmert [6] and of F. Grunewald and J. Schwermer [3], Theorem 5 implies

Theorem 6: *For all natural numbers $n \in \mathbf{N}$, there are surjective group homomorphisms*

$$\psi_n : G = \mathrm{SL}_2(\mathbf{Z}[x]) \longrightarrow F_n,$$

where F_n is the free, non-abelian group of rank n. The kernels ker ψ_n contain all unipotent elements of G.

The proofs of Theorems 4, 5, 6 will appear in three articles which are in preparation. In these articles, the work described above is put into a much broader framework.

References

[1] C.M. Campbell: 'Topics in the theory of groups', *Notes on Pure Mathematics*, 1 (1985), Department of Mathematics, Pusan National University.

[2] J.H. Conway: 'Advanced Problem' 5327, *Amer. Math. Monthly* 72(1965), 915.

[3] F. Grunewald and J. Schwermer: 'Free non-abelian quotients of SL(2) over integers of imaginary quadratic number fields', *J. Algebra*, 69(1981), 294-304.

[4] O.H. Kegel, A.C. Kim and J.L. Mennicke: 'On Fibonacci groups', *Proceedings of the Korea-Germany workshop 1987 on Algebra and Number Theory*, ed. by A.C. Kim and J.L. Mennicke, Jeung Moon Sa Publishing Co., Korea 1987, 163-167.

[5] M.F. Newman: 'The last of the Fibonacci groups', preprint.

[6] R. Zimmert: 'Zur SL$_2$ der ganzen Zahlen eines imaginär-quadratischen Zahlkörpers', *Inv. Math.*, 19(1973), 73-81.

Mathematische Fakultät
Universität Bielefeld
Postfach 8640, D 4800 Bielefeld 1
Federal Republic of Germany

PATHOLOGY IN THE REPRESENTATION THEORY OF INFINITE SOLUBLE GROUPS

Dedicated with love and respect
to my father for his eightieth birthday,
October 1989

Peter M. Neumann

1. Pathology.

These three lectures presented what was intended as an elementary exposition of some quite unfocussed mathematics. My purpose was to introduce an area where there are several openings for instructive research. In fact, one of the problems was, and is, to pin down just what 'pathology' should mean. I have not even attempted a proper definition; rather I have simply tried to indicate my intentions by means of a sequence of suggestive conjectures and examples.

Conjecture 1.1. For every countable group S there exists a finitely generated soluble group G such that $S \leq \mathrm{Aut}(G)$.

(Note that I use $\leq$ to mean either 'is isomorphic to a subgroup of', as here, or 'is a subgroup of'. The meaning should always be clear from the context.)

Conjecture 1.2. For every countable group S there exists a finitely generated soluble group G and there exists a finitely generated $\mathbf{Z}G$-module X such that $S \leq \mathrm{Aut}(X)$.

Conjecture 1.3. For every countable ring S (associative and with 1) there exists a finitely generated soluble group G and there exists a finitely generated $\mathbf{Z}G$-module X such that $S \leq \mathrm{End}(X)$.

Conjecture 1.4. For every pathological property $\mathcal{P}$ of modules there exists a finitely generated soluble group G and there exists a finitely generated $\mathbf{Z}G$-module X that has the property $\mathcal{P}$.

To make sense of this last conjecture we need to be able to recognise what a pathological property is. But, as I have already indicated, I do not wish to attempt precision. The following examples should show what I have in mind.

Example 1.5. The *non-Hopf* property. We say that a module X is non-Hopf if there exists a submodule Y such that $Y \neq \{0\}$ and $X/Y \cong X$. Equivalently, X is non-Hopf precisely when there is an endomorphism $\phi : X \longrightarrow X$ which is surjective but not injective.

Example 1.6. The *bad non-Hopf* property. We say that X is badly non-Hopf if $X \cong Y \oplus X_1$, with $Y \neq \{0\}$ and $X_1 \cong X$. Equivalently, X is badly non-Hopf precisely when there are endomorphisms $\phi, \psi : X \longrightarrow X$ such that $\psi\phi = 1$ and $\phi\psi \neq 1$.

Example 1.7. The *amoebic* property. We say that X is amoebic if $X \cong X \oplus X$ and $X \neq \{0\}$. Equivalently, X is amoebic precisely when there exist endomorphisms $\phi_1, \phi_2, \psi_1, \psi_2 : X \longrightarrow X$ such that $\psi_1\phi_1 = \psi_2\phi_2 = 1$ and $\phi_1\psi_1 + \phi_2\psi_2 = 1$.

Example 1.8. *Periodic pathology.* We shall say that X exhibits periodic pathology with period r if $m_1 X \cong m_2 X \iff m_1 \equiv m_2 \pmod{r}$, where mX denotes $X \oplus X \oplus \ldots \oplus X$ (m summands). Thus periodic pathology with period 1 is just the amoebic property. The conscientious reader is encouraged to find a way of expressing periodic pathology in terms of endomorphisms (in case of difficulties consult Corner [5]).

Example 1.9. The *Corner* property. We shall say that X has the Corner property if X is Hopfian (that is, not isomorphic to any proper factor module of itself) but there exists $W \neq \{0\}$ such that $X \oplus X \cong W \oplus X \oplus X$. Equivalently, X has the Corner property precisely when all surjective endomorphisms of X are automorphisms but there exist 2×2 matrices Φ, Ψ over $\text{End}(X)$ such that $\Psi\Phi = 1$ and $\Phi\Psi \neq 1$.

I hope it is clear by now that pathology should mean almost anything uncomfortable that I want it to, and that it will usually be expressible in terms of endomorphisms. At this point I should perhaps introduce a word of caution and a disclaimer. It may be that Conjecture 1.3 is false for relatively trivial reasons. I have not invested enough time in considering, for example, whether torsion in the additive group of S might in some cases provide good reason why S cannot be embeddable into the endomorphism ring of a finitely generated module over a finitely generated soluble group. I would be surprised if that were the case (after all, every S does occur as the endomorphism ring of a finitely generated module over some ring, namely, S is the endomorphism ring of the free module of rank 1 over the opposite ring S^{op}): but if so then Conjecture 1.3 should be modified by restricting S to lie in some civilised class of rings. The class of countable rings whose additive groups are torsion-free, or the class of countable rings whose additive groups are torsion-free and reduced (i.e. contain no divisible subgroups – compare [4]), or even the class of countable rings whose additive groups are free abelian, would prove enough pathology to convince most of us of the complexities of the representation theory of the groups in question.

To put these matters into context let us digress and consider what can happen in four other categories, those of sets, of abelian groups, of groups, and of G-spaces for a specified group G. In the category of SETS an object is finitely generated if and only if it is finite; thus the groups embeddable into automorphism groups of finitely generated objects are (by Cayley's Theorem) precisely the finite groups; assuming that the Axiom of Choice holds (as we always do), we find that an object is non-Hopf, badly non-Hopf or amoebic if and only if it is infinite; and neither periodic pathology with period $r \geq 2$ nor the Corner property can occur. In the category of ABELIAN GROUPS, or more generally the category MOD_R of right R-modules over any right-noetherian ring R, finitely generated objects satisfy the ascending chain condition, therefore are Hopfian and exhibit none of the pathologies described in 1.5 – 1.9 (for, if $\phi : X \longrightarrow X$ were a surjective endomorphism which was not injective then the submodules $\mathrm{Ker}(\phi^n)$ would form a strictly increasing chain, which is impossible); turning to non-finitely generated objects however, we see that non-Hopf, badly non-Hopf, even amoebic objects come very cheaply (any free abelian group of infinite rank is an example) and this is true in any non-trivial variety of algebraic systems; furthermore, A.L.S. Corner has described some wonderfully pathological examples of non-finitely generated countable abelian groups (see [4, 5, 6]): he has shown (1) that any countable ring whose additive group is torsion-free and reduced is isomorphic to $\mathrm{End}(G)$ for some countable reduced torsion-free abelian group G; (2) that for any $r \geq 2$ there is a countable torsion-free abelian group exhibiting periodic pathology modulo r; and (3) that there is a torsion-free abelian group having what I have called the Corner property (G is Hopfian but $G \oplus G$ is badly non-Hopf). In the category of all groups there is little more to be said. But what is of substantial interest is what happens for FINITELY GENERATED GROUPS. As far as automorphism groups go (compare Conjecture 1.2) if S is any countable group then, by a famous theorem of G. Higman, B.H. Neumann and Hanna Neumann, S is embeddable into some 2-generator group X and it is not hard to see that X can be chosen to have trivial centre so that conjugation gives an embedding of S into $\mathrm{Aut}(X)$. The question whether a finitely generated group could be isomorphic to a proper factor group of itself was formulated by H. Hopf nearly sixty years ago. He appears to have hoped and believed that this could not happen and accordingly terminology has become established in that direction: a group that is not isomorphic to any proper factor group of itself is known as a Hopf group. The first example of a finitely generated non-Hopf group did not come cheaply (See B.H. Neumann [20] and the commentary in [21], p.905), but by the early nineteen-sixties plenty more had come available. In the mid sixties my interest in varieties of groups led me to consider two questions about hopficity for finitely generated groups that generated an interest in hopficity for finitely generated modules; an analysis of the situation led me late in 1966 to the construction of Example 2.3 below; that example in turn suggested ideas for the manufacture of pathological finitely generated groups; Mary Tyrer (now Mary Jones) added greatly to those ideas in her Oxford D. Phil. thesis and in subsequent work (see [14, 15]) to construct (1) a finitely generated group isomorphic to a proper direct factor of itself; (2) a finitely generated group G such that $G \cong G \times G$; (3) a pair of finitely generated Hopf groups A, B such that $A \times B$ is isomorphic to a proper direct factor of itself; and (4) for any $r \geq 2$ a finitely generated group G such that $G^{m_1} \cong G^{m_2} \Longleftrightarrow m_1 \equiv m_2 \pmod{r}$, where G^m is the direct product $G \times G \times \cdots \times G$ with m factors. Of course (2), (3), (4) supersede (1), and (4) supersedes (2); a substantially simpler construction for (1) and (2) has been given by Meier [18].

The last category which is worth examining briefly for pathology is that of G-SPACES for a given group G, that is, sets with actions of G (by permutations). This is a slightly simpler category to understand than the category of $\mathbb{Z}G$-modules (i.e. abelian groups with actions of G by automorphisms) but, as we shall see later, what happens in the G-space category has a bearing on what happens in the module category. Traditional terminology for permutation groups can easily be correlated with the language of universal algebra. For example, a G-space Ω is finitely generated if and only if it has finitely many orbits and the cyclic (i.e. 1-generator) G-subspaces are precisely the orbits of G in Ω. Recall that a transitive G-space Ω is G-isomorphic to the coset space $(G : H)$ where H is a suitable subgroup of G, namely a stabiliser $G_\alpha := \{g \in G \mid \alpha g = \alpha\}$ for any α in Ω; and spaces $(G : H_1)$, $(G : H_2)$ are G-isomorphic if and only if H_1, H_2 are conjugate subgroups of G. It is perhaps not so familiar, but is easy to prove, that if Ω_1, Ω_2 are transitive G-spaces and $\alpha_1 \in \Omega_1$, $\alpha_2 \in \Omega_2$, then there is a G-morphism $\phi : \Omega_1 \longrightarrow \Omega_2$ such that $\phi : \alpha_1 \mapsto \alpha_2$ if and only if $G_{\alpha_1} \leq G_{\alpha_2}$. From this it is easy to determine what pathology can occur in the category of G-spaces. For example, any automorphism of a G-space Ω must permute the orbits among themselves and so we find that if Ω is finitely generated (i.e. has just finitely many orbits) then $\mathrm{Aut}(\Omega)$ is a direct product of finitely many groups (one for each isomorphism class of orbits), each of which is a wreath product $U \, \mathrm{wr} \, \mathrm{Sym}(m)$, where $\mathrm{Sym}(m)$ denotes the symmetric group of degree m and U denotes the automorphism group of a transitive G-space (namely, an orbit in Ω); furthermore, if that transitive G-space is isomorphic to the coset space $(G : H)$ then $U \cong N_G(H)/H$ (where $N_G(H) := \{g \in G \mid g^{-1}Hg = H\}$); in particular, if G is soluble and Ω is a finitely generated G-space then $\mathrm{Aut}(G)$ is soluble-by-finite and so the analogue of Conjecture 1.2 is very far from true for G-spaces. One can analyse the endomorphism semigroup of a G-space in a very similar way and again one finds that the analogue of Conjecture 1.3 is very far from true. As far as pathology and the analogue of Conjecture 1.4 go the situation is described in the next two theorems.

Theorem 1.10. *For a group G the following conditions are equivalent:*
  (1) *there exist $H \leq G$, $a \in G$ such that $H < a^{-1}Ha$;*
  (2) *there exists a transitive non-Hopf G-space;*
  (3) *there exists a finitely generated non-Hopf G-space.*

Proof. Suppose first that we have H, a as in (1). Define $\Omega := (G : H)$ and $\Omega' := (G : K)$ where $K := a^{-1}Ha$. Then Ω, Ω' are transitive G-spaces which are isomorphic because H, K are conjugate subgroups of G; on the other hand, the map $Hx \mapsto Kx$ is obviously a surjective G-morphism $\Omega \longrightarrow \Omega'$ which is not injective because any two cosets of H that lie in the same coset of K have the same image. Thus Ω is a transitive non-Hopf G-space. In fact, the formula $\phi : Hx \mapsto Hax$ describes a surjective endomorphism that is not injective. The converse, that (2) $\Longrightarrow$ (1), may be proved similarly: let $\phi : \Omega \longrightarrow \Omega$ be a surjective endomorphism of the transitive G-space Ω, let $H := G_\alpha$ for some α in Ω and let a be an element of G such that $\alpha\phi = \alpha a$. It is easy to check that $H \leq G_{\alpha\phi} = a^{-1}Ha$. Furthermore, if ϕ is not injective then $H < a^{-1}Ha$.

Obviously, (2) $\Longrightarrow$ (3). To prove the converse suppose that Ω is a finitely generated non-Hopf G-space and let $\phi : \Omega \longrightarrow \Omega$ be a surjective endomorphism that is not injective. If $\Omega_1, \cdots, \Omega_t$ are the G-orbits in Ω then, since $\Omega_i\phi$ is certainly a transitive G-space, $\Omega_1\phi, \cdots, \Omega_t\phi$ are orbits and since ϕ is surjective, they must be $\Omega_1, \cdots, \Omega_t$ in some order. This means that there is a permutation τ of $\{1, 2, \cdots, t\}$ such that $\Omega_i\phi = \Omega_{i\tau}$ for all i. Let n be the order of τ. Then ϕ^n is a surjective endomorphism of Ω that is not injective and $\Omega_i\phi^n = \Omega_i$ for all i. Consequently for some i the restriction of ϕ^n to Ω_i is a surjective endomorphism that is not injective, and so (3) $\Longrightarrow$ (2).

It is easy to produce examples of groups G with subgroups H conjugate to proper subgroups of themselves (I will exhibit some in §2 below) and so we get groups that have finitely generated non-Hopf G-spaces. Curiously, this is the worst pathology that can occur in this context:

Theorem 1.11. *For any group G every finitely generated G-space is weakly Hopf.*

(An object X is said to be weakly Hopf if it is not isomorphic to a proper direct factor of itself, that is, $X \cong X \times Y \implies |Y| = 1$).

Proof. Let Ω be a finitely generated G-space and suppose that $\Omega \cong \Gamma \times \Omega$ for some G-space Γ. Then also $\Omega \cong \Gamma \times \Gamma \times \Omega$. If $|\Gamma| > 1$ then there are at least 2 orbits in $\Gamma \times \Gamma$ and so if there are t orbits in Ω then there are at least $2t$ orbits in $\Gamma \times \Gamma \times \Omega$. This contradicts the fact that t is finite and $\Omega \cong \Gamma \times \Gamma \times \Omega$. Thus Ω is weakly Hopf.

We return now to the main theme of this paper and discuss some elementary points concerning Conjectures 1.1 to 1.4 and Examples 1.5 to 1.9. First, note that although I have chosen to express them as four separate conjectures 1.1 to 1.4 form a hierarchy.

Theorem 1.12. *If (A_i) denotes the assertion of Conjecture 1.i then $(A_4) \implies (A_3) \implies (A_2) \implies (A_1)$*

Proof. Let S be any countable group. Assuming (A_2) we have that there exists a finitely generated $\mathbf{Z}G$-module X with $S \le \mathrm{Aut}(X)$. Each $\mathbf{Z}G$-automorphism α of X induces an automorphism of the semi-direct product $X.G$ by the rule $xg \mapsto x^\alpha g$ and so $S \le \mathrm{Aut}(X.G)$. Since $X.G$ is a finitely generated soluble group we have (A_1). Thus $(A_2) \implies (A_1)$. That $(A_3) \implies (A_2)$ should be clear because we can take our countable ring in (A_3) to be the group ring $\mathbf{Z}S$ of the countable group S specified in (A_2). That $(A_4) \implies (A_3)$ is really no more than a matter of interpretation: for we should think of embeddability of an arbitrary countable ring into $\mathrm{End}(X)$ as being a manifestation of pathology.

Assertions (A_2) and (A_3) are both of the form $(\forall S)(\exists G)(\exists X)(S \le \mathrm{End}(X))$ and they each have strengthened versions $(A_2^!)$, $(A_3^!)$ of the form $(\exists G)(\forall S)(\exists X)(S \le \mathrm{End}(X))$. The extra strength is, however, more apparent than real.

Theorem 1.13. *If (A_2) holds then there is a finitely generated soluble group G such that for every countable group S there exists a finitely generated $\mathbf{Z}G$-module X such that $S \le \mathrm{Aut}(X)$. Similarly, if (A_3) holds then there is a finitely generated soluble group G such that for every countable ring S there exists a finitely generated $\mathbf{Z}G$-module X such that $S \le \mathrm{End}(X)$.*

Proof. Suppose that $(A_2^!)$ fails. Then if G_n is the free soluble group of length n and rank n then there is a countable group S_n that is not embeddable into $\mathrm{Aut}(X)$ for any finitely generated $\mathbf{Z}G_n$-module X. Let $S := S_1 \times S_2 \times S_3 \cdots$, so that S is a countable group into which every group S_n is embeddable. If there were a finitely generated soluble group G and a finitely generated $\mathbf{Z}G$-module X with $S \le \mathrm{Aut}(X)$ then, taking n larger than the soluble length and the number of generators of G, we would have that G was a quotient group of G_n and hence that X carried a natural structure as $\mathbf{Z}G_n$-module. This is impossible since $S_n \le S \le \mathrm{Aut}(X)$: thus G and X do not exist and so (A_2)

fails. Consequently $(A_2) \Longrightarrow (A_2^!)$. The proof of the second part of the theorem is almost exactly the same.

There is of course a similarly strengthened form $(A_4^!)$ of (A_4) and the implication $(A_4^!) \Longrightarrow (A_4)$ should follow from a version of the Compactness Theorem provided that we have a suitable understanding of what a pathological property is. This should be, in fact, one of the requirements to be satisfied by a definition of pathology. Notice that it follows from these implications $(A_i^!) \Longrightarrow (A_i)$ that if the conjectures are true then they hold with the added condition that G be of some specified soluble length. In fact I would add

Conjecture 1.14. Conjecture 1.1 is true with G soluble of length 4; Conjectures 1.2, 1.3, 1.4 hold with G soluble of length 3.

Let us say that the group G is *pathogenic* for the (pathological) property $\mathcal{P}$ if there exists a finitely generated $\mathbf{Z}G$-module X with property $\mathcal{P}$.

Theorem 1.15. (1) *If G has a pathogenic factor group then G is pathogenic.*
(2) *In general, if G has a pathogenic subgroup then G is pathogenic.*
(3) *If G is pathogenic and H is a subgroup of finite index then H is pathogenic.*

Proof. Part (1) is simply the fact (which we have already exploited in the proof of Theorem 1.12) that any module for a factor group of G may be viewed as a module for G. To prove (2) we need that if $H \leq G$ and X is a $\mathbf{Z}H$-module then the induced module $X^* := X \uparrow^G$ inherits most properties from X: if X is finitely generated then X^* is finitely generated as a $\mathbf{Z}G$-module; if $\phi \in \mathrm{End}_{\mathbf{Z}H}(X)$ then ϕ induces $\phi^* \in \mathrm{End}_{\mathbf{Z}G}(X^*)$ and the map $\phi \mapsto \phi^*$ gives an embedding $\mathrm{End}_{\mathbf{Z}H}(X) \leq \mathrm{End}_{\mathbf{Z}G}(X^*)$; furthermore, ϕ is injective (or surjective) if and only if ϕ^* is injective (or surjective respectively); if $X = X_1 \oplus X_2$ then $X^* = X_1^* \oplus X_2^*$; and so on. Part (3) is simply the fact that if X is a finitely generated $\mathbf{Z}G$-module and H is a subgroup of finite index in G then X is finitely generated as $\mathbf{Z}H$-module: if $x_1, \cdots, x_m$ are $\mathbf{Z}G$-generators for X and $G = t_1 H \cup \cdots \cup t_n H$ then $\{x_i t_j \mid 1 \leq i \leq m, \ 1 \leq j \leq n\}$ is a generating set for X as $\mathbf{Z}H$-module.

The words 'in general' that introduce part (2) of the theorem are intended to sound a cautionary note because there may be interesting difficulties in some cases. Suppose, for example, that $\mathcal{P}$ is periodic pathology with period 2. So we have a finitely generated $\mathbf{Z}G$-module X such that $X \cong X \oplus X \oplus X$ but $X \not\cong X \oplus X$. Then certainly $X^* \cong X^* \oplus X^* \oplus X^*$ but it seems just possible that if H is somehow badly embedded in G we might find that $X^* \cong X^* \oplus X^*$.

Question 1.16. Does there exist a group G with a subgroup H such that H has a finitely generated non-amoebic module X whose induced module $X \uparrow^G$ is amoebic?

[*Note added in proof, December 1988:* Dr. Annabelle McIver has shown me a positive answer to Question 1.16. She proves that if G is the group described in Example 2.6 below and H is the normal closure of c then there is a cyclic $\mathbf{Z}H$-module X, necessarily a Hopf module, such that $X \uparrow^G$

is amoebic.]

The statements of 1.1 to 1.4 include the specification that the group G should be finitely generated. This is of course the crux of the matter in Conjecture 1.1. But what matters for Conjecture 1.2, 1.3, 1.4 is that the module X should be finitely generated and I am not so concerned about finite generability of G in these contexts. As a matter of fact, there are many pathologies for which we have a "Local Theorem" of which the following is a typical example.

Theorem 1.17 (Djoković [7]; see also Bass [2], p.191, Orzech [24], and references quoted there). *If R is a ring with a finitely generated non-Hopf module then there is a finitely generated subring of R that has a finitely generated non-Hopf module.*

Proof. Let X be an R-module with a finite set $\{x_1, \cdots, x_n\}$ of generators and let $\phi : X \longrightarrow X$ be a surjective endomorphism that is not injective. Choose $y \in \mathrm{Ker}\phi$, $y \neq 0$. Since $x_1\phi, \cdots, x_n\phi$ lie in X they can be expressed as linear combinations of the given generators: thus there exist a_{ij} in R such that

$$(\alpha) \qquad\qquad x_j\phi = \sum_{i=1}^{n} x_i a_{ij} \qquad \text{for } 1 \leq j \leq n.$$

Likewise, since ϕ is surjective $x_1\phi, \cdots, x_n\phi$ generate X and so there exist b_{ij} in R such that

$$(\beta) \qquad\qquad x_j = \sum_{i=1}^{n} (x_i\phi) b_{ij} \qquad \text{for } 1 \leq j \leq n.$$

And there exist c_i in R such that

$$(\gamma) \qquad\qquad y = \sum_{i=1}^{n} x_i c_i.$$

Now let R_0 be the subring of R generated by all these coefficients a_{ij}, b_{ij}, c_i, let X_0 be the R_0-submodule of X generated by $x_1, \cdots, x_n$, and let ϕ_0 be the restriction of ϕ to X_0. The equations (α) tell us that ϕ_0 maps X_0 into X_0; equations (β) tell us that ϕ_0 is surjective as endomorphism of X_0; equation (γ) tells us that $y \in X_0$ and therefore that $\mathrm{Ker}(\phi_0) \neq \{0\}$. Thus X_0 is a finitely generated non-Hopf module for the finitely generated subring R_0 of R.

Corollary 1.18. *If G is a locally polycyclic-by-finite group and R is any commutative ring then finitely generated RG-modules are Hopfian.*

For, any finitely generated subring of RG is contained in a group ring R_0G_0 where R_0 is a finitely generated subring of R and G_0 is a finitely generated subgroup of G; then R_0 is noetherian (see, for example, [1], p.81) and G_0 is polycyclic-by-finite, and so R_0G_0 is right-noetherian by a famous theorem of P. Hall ([8], or [12], p.421). Consequently finitely generated R_0G_0-modules are noetherian, hence Hopfian, and so finitely generated RG-modules are Hopfian.

It should be clear how the argument for Theorem 1.17 may be adapted to prove the following similar local theorems.

Theorem 1.19. *A group G that is pathogenic for the pathological property $\mathcal{P}$ of modules has a finitely generated pathogenic subgroup in each of the following cases:*

(1) *$\mathcal{P}$ is the property of being finitely generated and badly non-Hopf;*

(2) *$\mathcal{P}$ is the property of being finitely generated and amoebic;*

(3) *$\mathcal{P}$ is the property of being finitely generated and having a specified finitely generated simple group embeddable into the automorphism group.*

Note that, since every countable group is embeddable into a finitely generated simple group (P. Hall [11] or [12], p.722; Meier [19]) case (3) includes some very complicated pathologies. The argument breaks down for periodic pathology and the Corner property, and it is not clear to me whether we have a local theorem in these cases or not.

Question 1.20. Does there exist a group (or ring) which is pathogenic in the sense that there exists a finitely generated module exhibiting periodic pathology or the Corner Property, but in which every finitely generated subgroup (or subring) is non-pathogenic?

Note that there is a local theorem also for the category of G-spaces.

Theorem 1.21. *There exists a finitely generated non-Hopf G-space if and only if there exists a finitely generated non-Hopf G_0-space for some finitely generated subgroup G_0 of G. Equivalently: there is a subgroup H conjugate to a proper supergroup of itself in G if and only if for some finitely generated subgroup G_0 of G there is a subgroup H_0 conjugate to a proper supergroup of itself in G_0.*

Finally, a few words about coefficient rings: I have spoken always about $\mathbf{Z}G$-modules; I could just as well have spoken of RG-modules for other commutative rings R; but it seems to me that the coefficient ring should not matter (so long as it is commutative) and I am encouraged in this opinion by Corollary 1.18. I shall continue to work with $\mathbf{Z}G$-modules (sometimes $\mathbf{Z}_p G$-modules, where $\mathbf{Z}_p$ is the ring of integers modulo p, perhaps thought of as $\mathbf{Z}G$-modules) throughout this paper.

2. Pathological examples.

As was mentioned earlier, it is very easy to find groups that have subgroups conjugate to proper supergroups of themselves. Since they will be used as ingredients in our later constructions we write down a couple of examples explicitly.

Example 2.1. If $G := \langle a, b \mid aba^{-1} = b^2 \rangle$ or if $G := \langle b \rangle$ wr $\langle a \rangle$ then G is finitely generated and metabelian (soluble of length 2), and has a subgroup H such that $H < a^{-1}Ha$.

The first example is the semi-direct product of the additive group $2^{-\infty}\mathbf{Z}$ consisting of all those rational numbers that have denominator a power of 2 by the infinite cyclic group $\langle a \rangle$ whose generator acts as multiplication by $\frac{1}{2}$; if $H := \langle b \rangle$ then obviously $H < a^{-1}Ha$. The second example may be presented as $\langle a, b \mid [a^{-i}ba^i, a^j ba^j] = 1$ for all $i, j \rangle$ (where $[x, y]$ denotes the commutator $x^{-1}y^{-1}xy$)

and if $H := \langle a^{-i}ba^i \mid i \leq 0 \rangle$ then $H < a^{-1}Ha$. In both groups the normal closure of $\langle b \rangle$ is abelian and has cyclic factor group.

Example 2.2. Let G be a group with a subgroup H conjugate to a proper supergroup of itself and let $X := 1_H \uparrow^G$ (the $\mathbf{Z}G$-module induced from the trivial $\mathbf{Z}H$-module $\mathbf{Z}$). Then X is a cyclic non-Hopf $\mathbf{Z}G$-module.

For, X is simply the permutation module $\mathbf{Z}\Omega$ where Ω is the coset space $(G : H)$. There is a surjective G-morphism $\Omega \longrightarrow \Omega$ that is not injective (see Theorem 1.10) and obviously this induces a surjective endomorphism of $\mathbf{Z}\Omega$ that is not injective.

Example 2.3 (See [22]): A finitely generated metabelian group G with a finitely generated badly non-Hopf module X.

Construction. Let U be a vector space with a basis $\{v_i\}_{i \in \mathbf{Z}}$ over $\mathbf{Z}_p$, where p is an odd prime number. Let a, b be the automorphisms of U defined by

$$a : v_i \mapsto v_{i+1} , \qquad b : v_i \mapsto \varepsilon_i v_i \qquad \text{for all } i,$$

where $\varepsilon_i = -1$ if $i = 0$, $\varepsilon_i = +1$ if $i \neq 0$. If $b_j := a^{-j}ba^j$, then

$$b_j : v_i \mapsto \begin{cases} -v_i & \text{if } i = j \\ v_i & \text{if } i \neq j \end{cases}$$

and so $\langle b_j \mid j \in \mathbf{Z} \rangle$ is an elementary abelian 2-group. Clearly this group $\langle b_j \mid j \in \mathbf{Z} \rangle$ is normalised by $\langle a \rangle$ and so if we define $G := \langle a, b \rangle \leq \text{Aut}(U)$ then G is metabelian, and in fact $G \cong \mathbf{Z}_2 \text{ wr } \mathbf{Z}$. Furthermore, U is naturally a $\mathbf{Z}_pG$-module (therefore also a $\mathbf{Z}G$-module) and it is cyclic, generated by u_0 (in fact, an easy calculation shows that U is a simple module, that is, it is generated by each of its non-zero members). Next we consider the cartesian power

$$P := U^{\mathbf{N}} := \{(u_0, u_1, u_2, \cdots) \mid u_i \in U\}.$$

This is a G-module with G acting componentwise. Let $\mathbf{x} := (v_0, v_1, v_2, \cdots) \in P$ and let $X := \langle \mathbf{x} \rangle_{\mathbf{Z}G}$, the $\mathbf{Z}G$-submodule of P generated by $\mathbf{x}$. By construction X is a cyclic module and our aim is to show that X is badly non-Hopf. To do this let

$$U_0 := \{(u_0, 0, 0, \cdots) \mid u_0 \in U\}, \qquad P_1 := \{0, u_1, u_2, \cdots) \mid u_i \in U\}$$

and $X_1 := X \cap P_1$. Now $\frac{1}{2}(\mathbf{x} - \mathbf{x}b) = (v_0, 0, 0, \cdots)$ and this generates U_0 as G-module. Thus $U_0 \leq X$ and since $P = U_0 \oplus P_1$ it follows that $X = U_0 \oplus (X \cap P_1) = U_0 \oplus X_1$. The point is that $X_1 \cong X$: for, if $\mathbf{x}_1 := (0, v_1, v_2, \cdots)$ then $\mathbf{x}_1$ certainly generates X_1 (because it is the projection of our generator of X into X_1); therefore also $\mathbf{x}_1 a^{-1}$ generates X_1, that is $(0, v_0, v_1, v_2, \cdots)$ generates X_1 and so $X_1 \cong X$. Thus X is isomorphic to a proper direct summand of itself, as required.

This construction can be adapted to prove the following assertion.

Theorem 2.4. *If G is a soluble group with a subgroup H conjugate to a proper supergroup of itself then there is a finitely generated badly non-Hopf $\mathbf{Z}G$-module.*

Proof. We may replace G by a factor group of itself that has minimal soluble length subject to having a subgroup conjugate to a proper supergroup of itself. Thus without loss of generality we may assume that G has subgroups H, K, N and an element a such that $H < K \leq N$, N is an abelian normal subgroup of G, and $K = a^{-1}Ha$. Choose $c \in K - H$ and let n be the order of c modulo H (so that $\langle c \rangle \cap H = \langle c^n \rangle$); note that $n = 0$ is a possibility). Let q be a prime divisor of n, let P be a prime number different from q, and let V be an irreducible $\mathbf{Z}_p\langle H, c \rangle$-module on which $\langle c \rangle \langle c^q \rangle$ acts faithfully (so that $\langle H, c^q \rangle$ is the kernel of the action of $\langle H, c \rangle$; the dimension of V will be the order of p modulo q). Define $U := V{\uparrow}^G$, the module induced from $\langle H, c \rangle$ to G. Let v be any non-zero element of V and let v_0 be the corresponding 'natural generator' of U: we may describe U as being (*qua* vector space) a direct sum $\bigoplus_{t \in T} V_t$ where T is a transversal for $\langle H, c \rangle$ in G, and then v_0 is the element corresponding to v in V_1. If $v_i := v_0 a^i$ for $i := 0, 1, 2, \cdots$ then

$$v_i c = v_0 a^i c = v_0 (a^i c a^{-i}) a^i,$$

and since $(a^i c a^{-i}) \in H$ for $i \geq 1$, we have that $v_i c = v_i$ for $i \geq 1$.

We shall find our badly non-Hopf module X inside the cartesian product $P := U^{\mathbf{N}}$. Indeed, let $\mathbf{x} := (v_0, v_1, v_2, \cdots) \in P$ and let $X := \langle \mathbf{x} \rangle_{\mathbf{Z}G}$. Now $\mathbf{x}c = (v_0 c, v_1, v_2, \cdots)$ and so $\mathbf{x} - \mathbf{x}c = (v_0(1 - c), 0, 0, \cdots)$. Since $v(1 - c)$ generates V as $\mathbf{Z}\langle H, c \rangle$-module $v_0(1 - c)$ generates U as $\mathbf{Z}G$-module. Thus, just as in the construction of Example 2.3, we have that $X = U_0 \oplus X_1$ where $U_0 := \{(u, 0, 0, \cdots) \mid u \in U\}$ and X_1 is generated by $(0, v_1, v_2, \cdots)$. Since $(0, v_1, v_2, \cdots)a^{-1} = (0, v_0, v_1, \cdots)$ we see that $X_1 \cong X$ and so X is badly non-Hopf, as required.

Question 2.5. Can the assumption that G be soluble be dropped from the statement of Theorem 2.4?

Example 2.6: A finitely generated soluble group of length 3 with a finitely generated amoebic module.

<u>Construction.</u> Let $I := 2^{-\infty}\mathbf{Z} := \{\frac{m}{2^r} \mid m \in \mathbf{Z}, r \in \mathbf{N}\}$ and let U be a vector space with basis $\{v_i\}_{i \in I}$ over $\mathbf{Z}_p$, where p is an odd prime number. Define three automorphisms a, b, c of U as follows:

$$a : v_i \mapsto v_{\frac{1}{2}i}, \qquad b : v_i \mapsto v_{i+1}, \qquad c : v_i \mapsto \varepsilon_i v_i,$$

for all i, where $\varepsilon_i = -1$ if $i \in \mathbf{N}$ and $\varepsilon_i = 1$ otherwise, and define $G := \langle a, b, c \rangle \leq \mathrm{Aut}(U)$. The automorphisms a, b act by permuting the specified basis of U and it is convenient to think of them as permutations of the index set I; Furthermore, $aba^{-1} : v_i \mapsto v_{i+2}$ and so $aba^{-1} = b^2$. Consequently, $\langle a, b \rangle$ is metabelian. If $g \in \langle a, b \rangle$ then $g^{-1}cg : v_i \mapsto \varepsilon_i' v_i$ where $\varepsilon_i' = -1$ if $i \in \mathbf{N}g$ and $\varepsilon_i' = 1$ otherwise. Therefore the normal closure of c in G is an elementary abelian 2-group and so G is soluble of length (at most) 3.

The amoebic module X that we are seeking is to be found as a submodule of the cartesian power

$$P := U^{\mathbf{N}} := \{(u_0, u_1, u_2, \cdots) \mid u_i \in U\}.$$

Let $\mathbf{x} := (v_0, v_1, v_2, \cdots)$ and let $X := \langle \mathbf{x} \rangle_{\mathbf{Z}G}$. Also let $\mathbf{x}_0 := (v_0, 0, v_1, 0, v_2, 0, \cdots)$, $X_0 := \langle \mathbf{x}_0 \rangle_{\mathbf{Z}G}$ and $\mathbf{x}_1 := (0, v_0, 0, v_1, 0, v_2, \cdots)$, $X_1 := \langle \mathbf{x}_1 \rangle_{\mathbf{Z}G}$. It is easy to check that $\mathbf{x}_0 = \frac{1}{2}(\mathbf{x}a - \mathbf{x}ac)$ and

$\mathbf{x}_1 = \frac{1}{2}(\mathbf{x}b^{-1}a - \mathbf{x}b^{-1}ac)$. Therefore, $\mathbf{x}_0$, $\mathbf{x}_1 \in X$ and so $X_0 \oplus X_1 \leq X$; but also $\mathbf{x} = \mathbf{x}_0 a^{-1} + \mathbf{x}_1 a^{-1}b$ and so $X = X_0 \oplus X_1$. On the other hand it is obvious that $X_0 \cong X_1 \cong X$ and so X is amoebic.

It seems to be very likely that if we modify the construction by taking $I := q^{-\infty}\mathbf{Z}$ and $a : v_i \mapsto v_{i/q}$, where $q \geq 3$, then we will get a cyclic module X that exhibits periodic pathology with period $q - 1$. It is this (and of course Example 2.6) that lead me to believe that any module pathology that occurs can already be found in modules for soluble groups of length 3, which is the substance of Conjecture 1.14.

Although it was not my primary purpose in this section to tackle Conjecture 1.2 (and hence Conjecture 1.1), the constructions do give some information about automorphisms because of the following theorem.

Theorem 2.7. (1) *If X is a badly non-Hopf module then $FS(\aleph_0) \leq \mathrm{Aut}(X)$, where $FS(\aleph_0)$ denotes the finitary symmetric group of countable degree.*
(2) *If X is amoebic and S is any countable locally finite group then $S \leq \mathrm{Aut}(X)$.*

Proof. (1) Since X is badly non-Hopf we have $X \cong Y_0 \oplus X_1$ where $Y_0 \neq \{0\}$ and $X_1 \cong X$. Thus we get that $X \cong Y_0 \oplus Y_1 \oplus \cdots \oplus Y_{n-1} \oplus X_n$ where $X_n \cong X$, and $Y_0 \cong Y_1 \cong \cdots \cong Y_{n-1}$. If f is any permutation of $\mathbf{N}$ of finite support, say a permutation that acts on $\{0, 1, \cdots, n-1\}$ and fixes all m with $m \geq n$, then f induces a module automorphism f^* of X that permutes the factors $Y_0, Y_1, \cdots, Y_{n-1}$ according to the permutation f and fixes X_n elementwise. This gives an embedding of the finitary symmetric group on $\mathbf{N}$, which is $FS(\aleph_0)$, into $\mathrm{Aut}(X)$.

(2) Since $X \cong X \oplus X$ we have $X \cong lX$ for every natural number l. Let $l_1 := 3$, let $l_{n+1} := l_n!$, and let L_1 be the symmetric group of degree 3 contained in $\mathrm{Aut}(X)$ and acting so as to permute the 6 factors in a direct sum decomposition $X \cong 6X$ regularly. Suppose now that we have L_{n-1} isomorphic to the symmetric group of degree l_{n-1}, contained in $\mathrm{Aut}(X)$ and acting so as to permute the summands in a direct sum decomposition $X \cong X_1 \oplus X_2 \oplus \cdots \oplus X_{l_n}$ regularly, where $X_i \cong X$ for all i. Decomposing each factor X_i into $(l_n - 1)!$ summands we find a subgroup L_n of $\mathrm{Aut}(X)$ such that $L_{n-1} \leq L_n$ and L_n is isomorphic to the symmetric group of degree l_n acting so as to permute regularly the factors in a decomposition $X \cong X_1 \oplus X_2 \oplus \cdots \oplus X_{l_n}$. Thus we get a chain $L_1 \leq L_2 \leq L_3 \leq \cdots$ of subgroups of $\mathrm{Aut}(X)$ whose union is the group C described by P. Hall ([10], or [12], p.567). This is the countable universal locally finite group; if S is any countable locally finite group then S is embeddable into C. Therefore any countable locally finite group S is embeddable into $\mathrm{Aut}(X)$.

Corollary 2.8. *There is a finitely generated soluble group G of length 4 such that any countable locally finite group S is embeddable into $\mathrm{Aut}(G)$.*

This comes from Example 2.6 and Theorem 2.7.(2) by the argument that proved $(A_2) \implies (A_1)$ in Theorem 1.12.

Theorem 2.7.(2) has been stated in a more restrained form than is necessary. What is really happening is that if X is any amoebic abject in any category then the group $G_{2,1}$ in G. Higman's notation [13] is embeddable into Aut(X) and C, the countable universal locally finite group, is embeddable into $G_{2,1}$. See Annabelle McIver's thesis [17] for details.

[*Note added in proof, December 1988.* Professor Brian Hartley has kindly drawn my attention to some further examples and conjectures concerning automorphism groups. In his paper [25] he proves that if G is an SI-group (that is, a group having a normal series with abelian factors) and X is a free $\mathbf{Z}G$-module of finite rank then Aut(X) has a normal SI-subgroup of finite index. By way of contrast he exhibits free kG-modules X of finite rank, where k is a finite field and G is a wreath product Z_p wr Z, such that Aut(X) has some distinctly unpleasant subgroups. In a letter dated 6 October 1988 he conjectures that for every countable group S there exists a finitely generated soluble group G such that S is embeddable into the group of units of kG for some field k, that is, $S \leq$ Aut(X) where X is the free cyclic kG-module.]

3. Lack of Pathology.

Many years ago, when I first got involved in this line of thinking, I believed that the greater the derived length of my soluble groups, the greater would be the possibilities for pathology in their module categories. Now, however, examples like those of §2 have persuaded me that as bad as is possible occurs already in the representation theory of soluble groups of length 3: hence Conjecture 1.14. Nevertheless, we should expect that for groups of small derived length or small nilpotent length levels of complication in the structure of the groups may be correlated with levels of pathology possible in their module categories. This is indeed so, as the following theorems indicate. The first, which is the heart of the matter, is due to Annabelle McIver.

Theorem 3.1 (McIver [16], [17]). *Let G be a finitely generated soluble group. If*

$$(\forall H \leq G)(\forall a \in G)(H \leq a^{-1}Ha \Longrightarrow H = a^{-1}Ha)$$

then G is polycyclic, and conversely.

Proof. Suppose the hypotheses hold, that is, no subgroup of G is conjugate to a proper supergroup of itself. We shall use induction on the soluble length d of G. Thus if $A := G^{(d-1)}$ (the last non-trivial term of the derived series of G) then $A \trianglelefteq G$, A is abelian, and, since our hypotheses hold for G/A which is of soluble length $d-1$, we may suppose that G/A is polycyclic. Therefore there is a series

$$A = G_0 < G_1 < G_2 < \cdots < G_n = G$$

in which $G_r \trianglelefteq G_{r+1}$ and G_{r+1}/G_r is cyclic for $r \leq n-1$; let g_{r+1} be a generator for G_{r+1} modulo G_r. Since G/A is polycyclic it is finitely presented and so A is finitely generated *qua* normal subgroup of G: let $a_1, \cdots, a_m$ be normal generators for A and let $A_r := \langle a_1, \cdots, a_m \rangle^{G_r}$ (the normal closure of $\{a_1, \cdots, a_m\}$ in G_r) for $r := 0, 1, \cdots, n$. We show by induction on r that A_r is a finitely generated abelian group.

Since A is abelian, $A_0 = \langle a_1, \cdots, a_m \rangle^A = \langle a_1, \cdots, a_m \rangle$ and so A_0 is finitely generated. Suppose therefore as inductive hypothesis that A_r is known to be finitely generated (it is certainly abelian, being a subgroup of A). Let

$$H := \langle g_{r+1} A_r g_{r+1}^{-1}, \; g_{r+1}^2 A_r g_{r+1}^{-2}, \cdots \rangle.$$

Then

$$g_{r+1}^{-1} H g_{r+1} = \langle A_r, \; g_{r+1} A_r g_{r+1}^{-1}, \cdots \rangle$$

and so $H \leq g_{r+1}^{-1} H g_{r+1}$. By assumption therefore $H = g_{r+1}^{-1} H g_{r+1}$ and so $A_r \leq H$. Since A_r is finitely generated there exists k such that

$$A_r \leq H_0 := \langle g_{r+1} A_r g_{r+1}^{-1}, \; g_{r+1}^2 A_r g_{r+1}^{-2}, \cdots, g_{r+1}^k A_r g_{r+1}^{-k} \rangle.$$

Then $H_0 \leq g_{r+1}^{-1} H_0 g_{r+1}$ and so $H_0 = g_{r+1}^{-1} H_0 g_{r+1}$, that is, H_0 is normalised by g_{r+1}. Furthermore, since G_r normalises A_r and is normalised by g_{r+1}, G_r normalises each group $g_{r+1}^i A_r g_{r+1}^{-i}$ and so G_r normalises H_0. Thus H_0 is normalised by G_{r+1}. It follows that $H_0 = A_{r+1}$ and, since H_0 is generated by finitely many finitely generated groups, that A_{r+1} is finitely generated. Ultimately therefore by induction we get that A is finitely generated as group and hence that G is polycyclic. The converse assertion follows immediately from the fact that polycyclic groups satisfy the ascending chain condition on subgroups.

McIver's theorem is the critical ingredient in the following theorem which gives a number of different characterisations of polycyclic groups in terms of lack of pathology in their representation theory.

Theorem 3.2. *Let G be a finitely generated soluble group. The following are equivalent:*
  (1) *G is polycyclic;*
  (2) *all finitely generated G-spaces are Hopfian;*
  (3) *all finitely generated $\mathbf{Z}G$-modules are Hopfian;*
  (4) *all finitely generated $\mathbf{Z}G$-modules are weakly Hopfian;*
  (5) *if X is a finitely generated $\mathbf{Z}G$-module then $\mathrm{Aut}(X)$ is residually finite;*
  (6) *up to isomorphism there are strictly fewer than $2^{\aleph_0}$ finitely generated $\mathbf{Z}G$-modules.*

Proof. In the light of Theorem 1.10, the equivalence of (1) and (2) is precisely McIver's theorem. If G is polycyclic then all finitely generated $\mathbf{Z}G$-modules satisfy the maximal condition on submodules (this is P. Hall's theorem cited in the proof of Corollary 1.18 above) and are therefore Hopfian: thus (1) $\Longrightarrow$ (3). That (3) $\Longrightarrow$ (4) is trivial; and (4) $\Longrightarrow$ (2) because if (2) fails then G has a subgroup conjugate to a proper supergroup of itself (Theorem 1.10) and so G has a finitely generated badly non-Hopf module (Theorem 2.4), that is, (4) fails. Thus (1), (2), (3), (4) are equivalent.

Suppose that G is polycyclic and X is a finitely generated $\mathbf{Z}G$-module. By a theorem of P. Hall (see [9], or [12], p.598) and J.E. Roseblade [23], X is residually finite. Therefore by a theorem of G. Baumslag [3], $\mathrm{Aut}(X)$ is residually finite. Conversely, suppose that G is not polycyclic. By (4) there exists a finitely generated badly non-Hopf module X and by Theorem 2.7 (1), $FS(\aleph_0) \leq \mathrm{Aut}(X)$; since $FS(\aleph_0)$ is not residually finite $\mathrm{Aut}(X)$ cannot be residually finite. Thus (5) is also equivalent to (1).

Again, if G is polycyclic then by P. Hall's theorem all free $\mathbf{Z}G$-modules Y of finite rank satisfy the maximal condition and so their submodules Z are finitely generated; consequently there are (up to isomorphism) only countably many possibilities for pairs Y, Z; but any finitely generated $\mathbf{Z}G$-module X is isomorphic to Y/Z for some free module Y of finite rank and submodule Z; therefore if G is polycyclic then there are only countably many finitely generated $\mathbf{Z}G$-modules up to isomorphism. If G is not polycyclic then by (4) we know that we have a finitely generated badly non-Hopf module X; thus $X = Y_0 \oplus X_1$ where $Y_0 \neq \{0\}$ and $X_1 \cong X$ and, splitting summands off one by one we see that

$$X = Y_0 \oplus Y_1 \oplus \cdots \oplus Y_{n-1} \oplus X_n$$

where $Y_i \neq \{0\}$ and $X_n \cong X$ for all i, n; the union of the submodules $Y_0 \oplus Y_1 \oplus \cdots \oplus Y_{n-1}$ is a submodule of X isomorphic to the infinite direct sum $Y_0 \oplus Y_1 \oplus \cdots$ and this obviously has $2^{\aleph_0}$ different submodules; consequently X has $2^{\aleph_0}$ different factor modules and since there are at most $\aleph_0$ homomorphisms of X to any given countable module X', hence at most $\aleph_0$ submodules Z of X such that $X' \cong X/Z$, our badly non-Hopf module X has $2^{\aleph_0}$ non-isomorphic factor modules. Thus also (6) is equivalent to (1).

Corollary 3.3. *If G is a locally soluble-by-finite group then the following are equivalent:*

(1) *G is locally polycyclic-by-finite;*

(2) *all finitely generated G-spaces are Hopfian;*

(3) *all finitely generated $\mathbf{Z}G$-modules are Hopfian;*

(4) *all finitely generated $\mathbf{Z}G$-modules are weakly Hopfian.*

This follows directly from Theorem 3.2 and the local theorems, 1.17, 1.19, 1.21.

The example of a finitely generated amoebic module described in 2.6 is a module for a soluble group of length 3, and a beautiful new example recently constructed by Annabelle McIver is a module for an abelian-by-polycyclic group, which, however, is also soluble of length 3. By way of contrast, I have long believed that no such construction is possible for metabelian groups (i.e. soluble groups of length 2):

Conjecture 3.4. There are no finitely generated amoebic $\mathbf{Z}G$-modules when G is met-abelian.

In fact it looks likely that the same should be true when G is metanilpotent (i.e. nilpotent-by-nilpotent). Annabelle McIver has made substantial progress towards a proof of the conjecture but it seems that much still remains to be done.

4. Postscript.

Most of the constructions described in this paper use as one of their main ingredients a subgroup H of our group G and an element a such that $H < a^{-1}Ha$. Obviously a must have infinite order. At the end of my lectures Otto Kegel asked what happens if we replace the universe of soluble (or locally soluble-by-finite) groups with the class of periodic groups. Then the main ingredient is not available because $H \leq a^{-1}Ha \implies H = a^{-1}Ha$. Of course the local theorems still hold and so we find that there is no pathology in the representation theory of locally finite groups. But Otto Kegel asks: do there exist finitely generated infinite periodic groups with finitely generated pathological modules? A very interesting question!

References

[1] M.F. Atiyah and I.G. Macdonald, *Introduction to commutative algebra*, Addison-Wesley, 1969.

[2] Hyman Bass, *Algebraic K-theory*, Benjamin, 1968.

[3] Gilbert Baumslag, 'Automorphism groups of residually finite groups', *J. London Math. Soc.*, 38(1963), 117-118.

[4] A.L.S. Corner, 'Every countable reduced torsion-free ring is an endomorphism ring', *Proc. London Math. Soc.* (3), 13(1963), 687-710.

[5] A.L.S. Corner, 'On a conjecture of Pierce concerning direct decompositions of abelian groups', *Proc. Coll. Ab. Groups, Tihany (Hungary) Sept. 1963*, Akad. Kiadó, Budapest 1964.

[6] A.L.S. Corner, 'Three examples on hopficity in torsion-free abelian groups', *Acta Math. Acad. Sci. Hung.*, 16(1965), 303-310.

[7] D.Ž. Djoković, 'Epimorphisms of modules which must be isomorphisms', *Canad. Math. Bull.*, 16(1973), 513-515.

[8] P. Hall, 'Finiteness conditions for soluble groups', *Proc. London Math. Soc.* (3), 4(1954), 419-436.

[9] P. Hall, 'On the finiteness of certain soluble groups', *Proc. London Math. Soc.* (3), 9(1959), 592-622.

[10] P. Hall, 'Some constructions for locally finite groups', *J. London Math. Soc.*, 34(1959), 305-319.

[11] P. Hall, 'On the embedding of a group in a join of given groups', *J. Austral. Math. Soc.*, 17(1974), 434-495.

[12] P. Hall: *Collected works of Philip Hall* (Compiled by K.W. Gruenberg and J.E. Roseblade), Clarendon Press, Oxford 1988.

[13] Graham Higman, *Finitely presented infinite simple groups*, Notes on Pure Mathematics 8, Australian National University 1974.

[14] J.M. Tyrer Jones, 'Direct products and the Hopf property', *J. Austral. Math. Soc.*, 17(1974), 174-196.

[15] J.M. Tyrer Jones, 'On isomorphisms of direct powers' in *Word Problems II* (ed. S.I. Adian, W.W. Boone and G. Higman), North-Holland 1980, 215-245.

[16] Annabelle McIver, 'Finitely generated non-Hopf modules', Submitted (June 1988) for publication.

[17] Annabelle McIver, *Finitely generated non-Hopf modules*. D. Phil. Thesis, Oxford 1988.

[18] David Meier, 'Non-Hopfian Groups', *J. London Math. Soc.* (2), 26(1985), 265-270.

[19] David Meier, 'Embeddings into simple free products', *Proc. Amer. Math. Soc.*, 93(1985), 387-392

[20] B.H. Neumann, 'A two-generator group isomorphic to a proper factor group', *J. London Math. Soc.*, 25(1950), 247-248.

[21] B.H. Neumann: *Selected works of B.H. Neumann and Hanna Neumann* (edited by D.S. Meek and R.G. Stanton), Charles Babbage Research Center, Winnipeg 1988.

[22] Peter M. Neumann, 'Endomorphisms of infinite soluble groups', *Bull. London Math. Soc.*, 12(1980), 13-16.

[23] J.E. Roseblade, 'Group rings of polycyclic groups', *J. Pure and Appl. Algebra*, 3(1973), 307-328.

[24] Morris Orzech, 'Onto endomorphisms are isomorphisms', *Amer. Math. Monthly*, 78(1971), 357-362.

[25] Brian Hartley, 'A conjecture of Bachmuth and Mochizuki on automorphisms of soluble groups', *Canad. J. Math.*, 28(1976), 1302-1310.

The Queen's College
Oxford OX1 4AW
England

23 September 1988.

A CHARACTERIZATION OF HEREDITARY P.I. RINGS[1]

Jae Keol Park and Klaus W. Roggenkamp

The results in this note arose more or less in discussions between both authors at a meeting in Pusan in the summer of 1988. The second author wishes to thank for the hospitality at that meeting.

Let Λ be a two-sided noetherian ring, satisfying a polynomial identity; i.e. Λ is P.I. We assume in addition that Λ is finitely generated over its center, and that Λ is prime. In this case the total ring of quotients of Λ is a simple K-algebra $A \simeq (D)_n$ for a skewfield D with centre K. Obviously $R = \Lambda \cap K$ is the centre of Λ, and we view Λ as an R-algebra, which is then finitely generated as R-module.

The aim of this note is to consider a question of E.P. Armendariz: Is the global dimension of a noetherian P.I. ring the supremum of the injective dimensions of the maximal two-sided ideals? In fact, the answer is positive in case the injective dimension of every maximal two-sided ideal is bounded by 1.

Theorem: *Let Λ be as above. Assume that for every maximal two-sided ideal $\mathfrak{M}$, inj.dim.$_\Lambda(\mathfrak{M}) \cdot \leq 1$, the injective dimension, as left Λ-module. Then Λ is hereditary.*

There is no loss, if we assume that R is not a field. We shall **reduce to the case, where R is complete local**: Λ is hereditary if and only if $R_{\hat{\mathfrak{p}}} \otimes_R \Lambda =: \Lambda_{\hat{\mathfrak{p}}}$ is hereditary for every maximal ideal $\mathfrak{p}$ of R. Here $R_{\hat{\mathfrak{p}}}$ is the completion of R at $\mathfrak{p}$ [RHD, Lemma 4.1] or [RMO, 3.24]. According to Bass' result [BA, 1.3] the condition on the injective dimension of the maximal two-sided ideals is inherited by the localizations at maximal ideals in R. Passing from the localization to the completion is faithfully flat [B, Pro.10, p.207]. Thus

we may assume that R is a complete local ring.

Lemma 1: *There does not exist a simple left Λ-module S with $\mathrm{Ext}^1_\Lambda(A, S) = 0$ for every finitely generated artinian Λ-module A.*

Proof. Let S be a simple left Λ-module with $\mathrm{Ext}^1_\Lambda(A, S) = 0$ for every finitely generated artinian left Λ-module A, and let P be the projective cover of S. We then have:

[1] This research was partially supported by the "Deutsche Forschungsgemeinschaft" and the "Korean Science and Engineering Foundation."

Claim 1: Let Q be any indecomposable projective; then $\mathrm{Hom}_\Lambda(P,Q) = 0$ for P not isomorphic to Q, and $\mathrm{End}_\Lambda(P)$ is a skewfield F.

Proof. Let $0 \neq \varphi \in \mathrm{Hom}_\Lambda(P,Q)$. Since Q is local, there exists an integer k such that $\mathrm{Im}(\varphi) \nsubseteq \mathrm{rad}(\Lambda)^k$, pick k smallest possible. Note that Λ is also complete in the $\mathrm{rad}(\Lambda)$-adic topology, Λ being finitely generated over R, and so $\bigcap_i \mathrm{rad}(\Lambda)^i = 0$. Then S lies in the socle of $Q/\mathrm{rad}(\Lambda)^k \cdot Q$; However, S is injective with respect to finitely generated artinian Λ-modules, and $Q/\mathrm{rad}(\Lambda)^k$ is such a module. Hence S is a direct summand of $Q/\mathrm{rad}(\Lambda)^k \cdot Q$; i.e. $P \simeq Q$ and φ must be an isomorphism.

Thus Λ has the following form – We write (X,Y) for $\mathrm{Hom}_\Lambda(X,Y)$:

$$\Lambda = \left[\begin{array}{cc} (X,X) & (X,Y) \\ 0 & (Y,Y) \end{array} \right],$$

where $X = \bigoplus\limits_{Q_j \neq P} Q_j^{(n_j)}$ and $Y = \bigoplus\limits_{j=1}^{t} P^{(n_j)}$, and ${}_\Lambda\Lambda \simeq X \oplus Y$. But then (X,Y) is a two-sided nilpotent ideal, a contradiction, since A, the total ring of quotients of Λ is simple. This completes the proof of Lemma 1.

Because of Lemma 1 **we may assume** that Λ does not have a simple module S with $\mathrm{Ext}_\Lambda^1(A,S) = 0$ for every artinian finitely generated Λ-module A.

Lemma 2: *Let $\mathfrak{M}$ be a maximal two-sided ideal in Λ such that – as left Λ-module – inj.dim.${}_\Lambda(\mathfrak{M}) \leq 1$. Then $\mathrm{E}(\Lambda) \neq \mathrm{E}(\mathfrak{M})$ and so Λ injects into $\mathrm{E}(\Lambda)$; moreover, $\mathrm{soc}_\Lambda(\mathrm{E}(\Lambda)/\Lambda) \neq 0$. Here $\mathrm{E}(X)$ denotes the injective envelope of the left Λ-module X, and $\mathrm{soc}_\Lambda(X)$ is the socle of X.*

Before we come to the proof we shall recall the **Goldie-dimension** of a module: For a left Λ-module M the Goldie-dimension of M is defined as the largest integer n such that

$$M \supset X_1 \oplus X_2 \oplus \cdots \oplus X_n,$$

where X_i are non-trivial Λ-submodules, or infinity.

We **recall** some facts about Goldie-dimensions [G]:

1.) If Λ is left noetherian, then the Goldie-dimension of every finitely generated Λ-module is finite.

2.) The Goldie-dimension is additive w.r.t. direct sums.

3.) X has Goldie-dimension zero iff X is zero.

4.) X and $\mathrm{E}(X)$ have the same Goldie-dimension.

We shall use the Goldie-dimension as a "rank-function".

Proof of Lemma 2: We first show that Λ injects into $\mathrm{E}(\mathfrak{M})$. Let $\mathrm{E} = \mathrm{E}(\mathfrak{M})$. Since inj.dim.${}_\Lambda(\mathfrak{M}) \leq 1$, we have an exact sequence

$$0 \longrightarrow \mathfrak{M} \longrightarrow E \longrightarrow E/\mathfrak{M} \longrightarrow 0, \tag{1}$$

where $E/\mathfrak{M}$ is an injective Λ-module. The injection $\mathfrak{M} \longrightarrow \Lambda$ induces a map $\rho : \Lambda \longrightarrow E$.

Claim 2: The map ρ is injective.

Proof. In fact $\mathrm{Ker}(\rho) \cap \mathfrak{M} = 0$. Hence $\mathrm{Ker}(\rho)$ injects into $\Lambda/\mathfrak{M} = S^{(n)}$, where S is a simple Λ-module. Thus $\mathrm{Ker}(\rho)$ is semi-simple. But then S is projective, since the map

$$\mathrm{Ker}(\rho) \longrightarrow \Lambda \longrightarrow \Lambda/\mathfrak{M}$$

induces a splitting. Thus the sequence of two-sided Λ-modules

$$0 \longrightarrow \mathfrak{M} \overset{\varphi}{\longrightarrow} \Lambda \overset{\psi}{\longrightarrow} \Lambda/\mathfrak{M} \longrightarrow 0$$

is split as sequence of left Λ-modules. However, then it is also two-sided split: Let $\kappa : \Lambda/\mathfrak{M} \longrightarrow \Lambda$ be a left splitting; then $\psi\kappa \in \mathrm{End}(_\Lambda\Lambda) = \Lambda$ is an idempotent (we write maps on the right) and so $\eta = 1 - \psi\kappa$ is an idempotent in Λ with $\mathfrak{M} = \Lambda\eta$ a two-sided ideal; i.e., $\eta\Lambda \subset \Lambda\eta$. Let $\epsilon = 1 - \eta$, then

$$\Lambda = \Lambda\eta \oplus \Lambda\epsilon = \eta\Lambda \oplus \epsilon\Lambda.$$

Since $\eta\Lambda \subset \Lambda\eta$, we have $\Lambda\epsilon \subset \epsilon\Lambda$, and so $\epsilon\Lambda\eta = 0 = \eta\Lambda\epsilon$, and so the Peirce decomposition shows that ϵ is central, and so $\Lambda = \Lambda_0 \oplus \Lambda_1$, but then Λ is indecomposable as ring. This proves Claim 2.

Hence we may assume that the map $\rho : \Lambda \longrightarrow E$ is injective.

Claim 3: $\mathrm{E}(\mathfrak{M}) \simeq \mathrm{E}(\Lambda)$.

Proof. The injection $\mathfrak{M} \longrightarrow \Lambda$ shows that $\mathrm{E}(\mathfrak{M})|\mathrm{E}(\Lambda) - X|Y$ means that X is a direct summand of Y. Since by Claim 2 Λ injects into $\mathrm{E}(\mathfrak{M})$, we conclude that $\mathrm{E}(\Lambda)|\mathrm{E}(\mathfrak{M})$. Using the above quoted properties of the Goldie-dimension, we conclude that $\mathrm{E}(\Lambda) \simeq \mathrm{E}(\mathfrak{M})$.

Claim 4: There exists a non-split extension

$$0 \longrightarrow S_i \overset{\alpha}{\longrightarrow} X'_i \longrightarrow T_i \longrightarrow 0, \tag{2}$$

where S_i is a simple constituent of $\Lambda/\mathfrak{M}_i$, and T_i is a simple Λ-module.

Proof. $\qquad\qquad\qquad \bigcap_i \mathfrak{M}_i = \mathrm{rad}(\Lambda)$

is the Jacobson-radical of Λ, and $\Lambda/\mathrm{rad}(\Lambda)$ is semi-simple. Since Λ is noetherian, $\mathrm{rad}(\Lambda)^2 \neq \mathrm{rad}(\Lambda)$, and so $\Lambda/\mathrm{rad}(\Lambda)^2$ is not semi-simple. Let S_i be defined by $\Lambda/\quad_i = S_i^{(n_i)}$. If S_i does inject into $\mathrm{rad}(\Lambda)/\mathrm{rad}(\Lambda)^2$, we are done. However, there always exists –according to Lemma 1 - a finitely generated artinian left Λ-module A and a non-split extension

$$0 \longrightarrow S_i \longrightarrow X' \longrightarrow A \longrightarrow 0,$$

and hence $S_i \subset \mathrm{rad}_\Lambda(X')$. Let j be largest such that $S_i \subset \mathrm{rad}_\Lambda(X')^j$, then S_i lies in the socle of

$$\mathrm{rad}_\Lambda(X')^{j-1}/\mathrm{rad}_\Lambda(X')^{j+1},$$

and it also lies in $\mathrm{rad}_\Lambda(X')^j$. Thus we can construct an exact sequence (2).

Definition: Let $X_i = \overset{n_i}{\oplus} X'_i$, where X'_i is defined in (2). Then we have the exact sequence

$$0 \longrightarrow S_i^{(n_i)} \overset{\alpha}{\longrightarrow} X_i \longrightarrow T_i^{(n_i)} \longrightarrow 0. \tag{3}$$

Because of the assumption on the injective dimension of maximal two-sided ideals we also have the exact sequence

$$0 \longrightarrow \mathfrak{M}_i \longrightarrow E(\Lambda) \longrightarrow E(\Lambda)/\mathfrak{M}_i \longrightarrow 0,$$

and hence the following exact sequence, which can be extended by β

$$0 \longrightarrow \quad S_i^{(n_i)} \quad \longrightarrow \quad E(\Lambda)/\mathfrak{M}_i \quad \longrightarrow E(\Lambda)/\Lambda \longrightarrow 0, \tag{4}$$

$$\alpha \downarrow \quad \nearrow$$

$$X_i^{(n_i)} \quad \beta$$

since $E(\Lambda)/\mathfrak{M}_i$ is injective. The map β surely is injective, and thus we have an injection $X_i^{(n_i)}/S_i^{(n_i)} \simeq T_i^{(n_i)}$ into $E(\Lambda)/\Lambda$, which therefore has a non-trivial socle. This completes the proof of Lemma 2.

Observation: In general, there is no reason, why $\mathrm{soc}_\Lambda(E(\Lambda)/\Lambda)$ is finitely generated. However, in later applications we need finitely generated modules. We now pick a special finitely generated submodule in $\mathrm{soc}_\Lambda(E(\Lambda)/\Lambda)$: In (4) we had an injection of X_i into $E(\Lambda)/\mathfrak{M}_i$, where X_i was constructed above in (3). Let $X = \oplus_i X_i \subset \oplus_i E(\Lambda)/\mathfrak{M}_i$. Then we have a natural injection of $\Lambda/\mathrm{rad}(\Lambda)$ into $\oplus_i E(\Lambda)/\mathfrak{M}_i$.

Definition: Let Y be the inverse image in $E(\Lambda)$ of the submodule X in $\oplus_i E(\Lambda)/\mathfrak{M}_i$.

Lemma 3: $_\Lambda \mathrm{rad}(\Lambda)$ *is a generator as left Λ-module.*

Proof. We first show: $\mathrm{rad}(\Lambda) \cdot Y = \Lambda$.
Because of the construction, and since $\mathrm{inj.dim.}_\Lambda(\mathfrak{M}_i) \leq 1$, and so $E(\Lambda)/\mathfrak{M}_i$ is injective, and $\Lambda/\mathrm{rad}(\Lambda)$ injects into $\oplus_i E(\Lambda)/\mathfrak{M}_i$. Moreover

$$\mathrm{soc}_\Lambda(X) \simeq \Lambda/\mathrm{rad}(\Lambda) \simeq \mathrm{rad}_\Lambda(X);$$

in fact, since $\mathrm{rad}_\Lambda(\Lambda)^2 \cdot X = 0$, surely $\mathrm{rad}_\Lambda(X) \subset \mathrm{soc}_\Lambda(X)$. If this inclusion were proper, then there would exist a simple module T in the socle of $E(\Lambda)/\Lambda$, which lies also in the socle of X, but this can not happen, because of the construction of X, a contradiction. Since the kernel of the natural map $Y \longrightarrow X$ lies in $\mathrm{rad}(\Lambda) \cdot Y$, we conclude $\mathrm{rad}(\Lambda) \cdot Y = \mathrm{rad}_\Lambda(Y)$, Y being finitely generated, and hence $\mathrm{rad}(\Lambda) \cdot Y = \Lambda$.

In order to show that $\mathrm{rad}(\Lambda)$ is a generator, we note that the equation,

$$\mathrm{rad}(\Lambda) \cdot Y = \Lambda$$

allows us to interpret the elements in Y as elements in

$$\mathrm{Hom}_\Lambda(_\Lambda\mathrm{rad}(\Lambda),_\Lambda \Lambda).$$

But then the standard arguments show that $\mathrm{rad}(\Lambda)$ is a generator, i.e. Λ is isomorphic to a direct summand of a direct sum of some copies of $\mathrm{rad}(\Lambda)$ as left Λ-module.

Claim 5: $\mathrm{rad}(\Lambda)$ *is projective as left Λ-module.*

Proof. According to the above, $\mathrm{rad}(\Lambda)$ is a generator. Let

$$_{\Lambda}\Lambda \simeq \overset{n}{\underset{i=1}{\oplus}} P_i^{(n_i)}$$

be the direct decomposition of Λ into indecomposable projectives. We <u>note</u> that

$$\text{Goldie-dimension}(P_i) = \text{Goldie-dimension}(\mathrm{rad}_{\Lambda}(P_i)),$$

provided, $\mathrm{rad}_{\Lambda}(P_i) \neq 0$.

<u>In fact</u>, let $n = \text{Goldie-dimension}(P_i)$, and let

$$P_i \supset X_1 \oplus X_2 \oplus \cdots \oplus X_n.$$

Assume $X_1 \oplus X_2 \oplus \cdots \oplus X_n$ is not contained in $\mathrm{rad}_{\Lambda}(P_i)$, then there exists an i such that $X_i \nsubseteq \mathrm{rad}_{\Lambda}(P_i)$. However, P_i is local – $P_i/\mathrm{rad}_{\Lambda}(P_i)$ ia simple – and so $P_i = \mathrm{rad}_{\Lambda}(P_i) + X_i$. Nakayama's lemma then shows that $X_i = P_i$, and so Goldie-dimension$(P_i) = 1$. Since $\mathrm{rad}_{\Lambda}(P_i) \neq 0$, we conclude that Goldie-dimension$(\mathrm{rad}_{\Lambda}(P_i)) = 1$. Whence we have the statement.

We are now in the position to show that rad(Λ) is projective, even that it is a progenerator:

Assume that the P_i's are numbered with respect to decreasing Goldie-dimension. So P_1 has maximal Goldie-dimension. Since $\mathrm{rad}(\Lambda)$ is a generator – as left Λ-module – we conclude that

$$P_1 \oplus X \simeq \mathrm{rad}_{\Lambda}(P_i) \quad \text{for some } i.$$

Hence we have that

$$\text{Goldie-dimension}(P_1) + \text{Goldie-dimension}(X) =$$
$$\text{Goldie-dimension}(\mathrm{rad}_{\Lambda}(P_i)) = \text{Goldie-dimension}(P_i),$$

since $\mathrm{rad}_{\Lambda}(P_i) \neq 0$, Λ being indecomposable as ring, we conclude that $X = 0$, since P_1 has maximal Goldie-dimension. Moreover, Goldie-dimension$(P_1) = $ Goldie-dimension(P_i). So we may assume $i = 2$. Continuing this way, we eventually reach the module P_k of the same Goldie-dimension as P_1 such that $\mathrm{rad}_{\Lambda}(P_1) = P_k$. We now look at multiplicities:

$$n_1 \leq \mu_k(\mathrm{rad}(\Lambda)), \quad n_2 \leq \mu_1(\mathrm{rad}(\Lambda)), \quad \cdots, \quad n_k \leq \mu_{k-1}(\mathrm{rad}(\Lambda)),$$

where $\mu_i(\mathrm{rad}(\Lambda))$ is the multiplicity of P_i as a direct summand of $\mathrm{rad}(\Lambda)$. But this can only happen if $\mathrm{rad}_{\Lambda}(P_1^{(n_1)} \oplus \cdots \oplus P_k^{(n_k)})$ is projective. Now continue in the same way with P_{k+1}. This shows that $\mathrm{rad}(\Lambda)$ is projective.

Claim 6: Λ is hereditary.

Proof. We shall show that submodules of projective Λ-modules are projective. So let P_i, $1 \leq i \leq t$ be the non-isomorphic indecomposable projective left Λ-modules. Let $X \subset \oplus_i P_i^{(n_i)}$ be a Λ-submodule. Using induction on $\sum n_i = n$ we shall show that X is projective.

In case $n = 1$, we note that $\mathrm{rad}_\Lambda(P_i)$ is projective indecomposable: In fact, $\Lambda \simeq \mathrm{rad}(\Lambda)$, so $\mathrm{rad}(\Lambda)$ has no multiplicities, and so $\mathrm{rad}(\Lambda) \simeq P_j \oplus P_k \oplus \cdots$ can not occur. Consequently, $\mathrm{rad}_\Lambda(P_i) \simeq P_{\sigma(i)}$, where σ is a permutation of $\{1, \cdots, t\}$. Hence the lattice of submodules of P_i is linear and consists of projective modules only.

If $n > 1$, then let $\rho : \oplus_i P_i^{(n_i)} \longrightarrow P_{i_0}$ be surjection. If $\rho|_X = 0$, then we are done by induction. Otherwise $\mathrm{Im}(\rho|_X)$ is projective by the above, and so it is a direct summand of X. Since $\mathrm{Ker}(\rho|_X)$ is projective by induction, we are done. This completes the proof of the theorem.

References

[A]   Armendariz, E.P. Private communication 1988.

[BA]   Bass, H. 'Injective dimension in noetherian rings', *Trans. Amer. Math. Soc.* 102(1962), 18-29.

[B]   Bourbaki, N. *Commutative Algebra*, Hermann, Paris 1972.

[G]   Goodearl, K.R. *Ring Theory – Nonsingular Rings and Modules*, Marcel Dekker, New York, 1976.

[RHD]   Roggenkamp, K.W. – V. Huber–Dyson, *Lattices over Orders I*, Springer Lecture Notes in Math. 142, 1969.

[RMO]   Reiner, I. *Maximal Orders*, Academic Press, New York, 1975.

Jae Keol Park
Department of Mathematics
Pusan National University
Pusan, 609–735
Republic of Korea

Klaus W. Roggenkamp
Mathematische Institut
Universität Stuttgart
7000 Stuttgart 80
Federal Republic of Germany

SOME PROBLEMS IN COMBINATORIAL GROUP THEORY

Stephen J. Pride

0. Introduction.

The main aim of this article is to present a list of (for the most part new) open problems in combinatorial group theory. However, I have also included a fair amount of factual material, partly for interest's sake, and partly as motivation for the problems.

I thank Peter Kropholler for some useful discussions in connection with this article.

1. Infinite Coxeter Groups.

A *graph* Γ consists of a set $\mathbf{x}$ (the *vertices*), and a set $\mathbf{e}$ of two-element subsets of $\mathbf{x}$ (the *edges*). We assume for convenience that Γ is finite and has more than one vertex.

Consider a pair (Γ, ϕ) where Γ is a graph as above, and ϕ is a mapping from $\mathbf{e}$ to $\{2, 3, 4, \cdots\}$. The *Coxeter group* $C(\Gamma, \phi)$ associated with this pair is the group given by the presentation

$$\langle \mathbf{x} \; ; \; x^2 \; (x \in \mathbf{x}), \; (xy)^{\phi\{x,y\}} \; (\{x, y\} \in \mathbf{e}) \rangle.$$

Necessary and sufficient conditions for $C(\Gamma, \phi)$ to be finite are well-known ([4, pp. 193-194] or [28, pp.359-360]). (These conditions are usually expressed, not in terms of Γ, but in terms of another graph called the *Coxeter graph*) .

Appel and Schupp [1], [2], defined a Coxeter group to be of *large type* if $2 \notin \operatorname{Im} \phi$. In recent work [26], R. Stöhr and I considered a wider class of Coxeter groups, which we called *aspherical*. A Coxeter group $C(\Gamma, \phi)$ is said to be *aspherical* if any three distinct elements of $\mathbf{x}$ generate an infinite subgroup of $C(\Gamma, \phi)$.

An alternative characterization of asphericity is the following: for any three edges e_1, e_2, e_3 of Γ which form a triangle

(*) $\qquad\qquad\qquad \frac{1}{\phi(e_1)} + \frac{1}{\phi(e_2)} + \frac{1}{\phi(e_3)} \leq 1.$

In [7], Davis associates a simplicial complex $K_0(\Gamma, \phi)$ with the pair (Γ, ϕ) as follows: the simplices of $K_0(\Gamma, \phi)$ are the non-empty subsets $\mathbf{y}$ of $\mathbf{x}$ such that the subgroup of $C(\Gamma, \phi)$ generated by $\mathbf{y}$ is *finite*. Note then that $C(\Gamma, \phi)$ is aspherical if and only if dim $K_0(\Gamma, \phi) \le 1$.

Problem 1. What is the virtual cohomological dimension (vcd) of $C(\Gamma, \phi)$?

It is shown in [7, Proposition 14.1] that vcd $C(\Gamma, \phi) \le \dim K_0(\Gamma, \phi) + 1$. Thus, in particular, if $C(\Gamma, \phi)$ is aspherical then vcd $C(\Gamma, \phi) \le 2$. An alternative proof of this latter fact can be found in [26, Theorem 1, Corollary 2]. For aspherical Coxeter groups one can show that:

$$\text{vcd } C(\Gamma, \phi) = \begin{cases} 1 & \text{if } \Gamma \text{ is a forest} \\ 2 & \text{otherwise.} \end{cases}$$

(See [26, Theorem 1, Corollary 3].)

Problem 2. Solve the isomorphism problem for Coxeter groups.

The most optimistic way of tackling this would be to try to show that the following problem has an affirmative answer.

Problem 2′. If $C(\Gamma, \phi)$ is isomorphic to $C(\Gamma', \phi')$ then is there a graph isomorphism $\theta : \Gamma \longrightarrow \Gamma'$ such that $\phi'\theta(e) = \phi(e)$ for all edges e of Γ?

J. Tits has informed me that he has made some progress on the isomorphism problem for Coxeter groups.

Problem 3. (B.H. Neumann) When are Coxeter groups SQ-universal?

Recall that a group G is *SQ-universal* if every countable group is isomorphic to a subgroup of a quotient of G.

Very recently, Eric Fennessey has shown that, with the exception of the group $A = \langle a, b, c \; ; \; a^2, b^2, c^2, (ab)^3, (bc)^3, (ca)^3 \rangle$, Coxeter groups of large type on at least three generators are SQ-universal.

2. Residual finiteness of small cancellation groups.

By a *small cancellation group* we will mean a group given by a presentation satisfying one of the three basic conditions C(6), C(4) and T(4), or C(3) and T(6) (see [16, pp.241-242] for definitions).

It is easy to construct non-trivial 2-generator infinitely related small cancellation groups which are not residually finite. In fact, we can construct such groups having no proper subgroups of finite index. For each positive integer n let U_n, V_n be words in a^n and b^n, and let

$$G = \langle a, b \; ; \; aU_1, bV_1, aU_2, bV_2, aU_3, bV_3, \cdots \rangle.$$

By choosing the U's and V's judiciously we can arrange for G to be a non-trivial small cancellation group. Now if M is a subgroup of finite index in G then M must contain the normal closure of $\{a^n, b^n\}$ for some n. But putting $a^n = b^n = 1$ in G gives the trivial group, so $M = G$.

Problem 4. Let G be a *finitely presented* small cancellation group. Does G have a proper subgroup of finite index?

It has been asked (see for example [27], [29, Problem F.5]) whether finitely presented small cancellation groups are residually finite. In view of the fact that Problem 4 remains unresolved, this question seems rather optimistic. However, one could test it out by looking at some specific small cancellation groups. An interesting class of small cancellation groups which might be amenable to investigation is the following.

Consider a presentation

$$(+) \qquad\qquad \langle s_1, \mathbf{x} \ ; \ s_1 U s_1^{-1} V^{-1} \rangle$$

where U and V are non-empty, cyclically reduced words on $\mathbf{x}$ of the same length. Write $U = u_1 u_2 \cdots u_n$, $V = v_1 v_2 \cdots v_n$ where the u's and v's belong to $\mathbf{x} \cup \mathbf{x}^{-1}$. Then the presentation is Tietze-equivalent to the C(3), T(6)-presentation

$$\langle s_1, t_1, \cdots, s_n, t_n, \mathbf{x} \ ; \ s_i u_i t_i^{-1}, t_i^{-1} v_i s_{i+1} \ (1 \le i \le n) \rangle$$

where the subscripts are computed mod n.

Now many one-relator groups with a presentation of the form $\langle t, \mathbf{x} \ ; \ tWt^{-1} = Z \rangle$ (W, Z words on $\mathbf{x}$) are not residually finite [18], but *can this happen if W and Z are cyclically reduced and of the same length?*

Problem 5. If G has a presentation as in $(+)$ then is G residually finite?

3. Generalized conjugacy problems.

Everyone is familiar with the word problem and conjugacy problem for groups given by presentations. From a geometric point of view it is a bit unnatural to consider these problems in isolation, and so we introduce an infinite family of problems, the first two of which are the word problem and conjugacy problem.

Let a group G be given by a presentation $P = \langle \mathbf{x}; \mathbf{r} \rangle$ where we assume that the elements of $\mathbf{r}$ are non-empty and cyclically reduced, and that if $R \in \mathbf{r}$ then no cyclic permutation of $R^{\pm 1}$ other than R itself belongs to $\mathbf{r}$.

Let $(W_1, W_2, \cdots, W_k)$ be a finite sequence of non-empty cyclically reduced words in the generating symbols of G. We assume that no W_i is a cyclic permutation of a relator or the inverse of a relator. We will say that this sequence is *dependent* if there are words $U_1, U_2, \cdots, U_k$ such that $U_1 W_1 U_1^{-1} U_2 W_2 U_2^{-1} \cdots U_k W_k U_k^{-1} = 1$ in G.

If m is a positive integer then the *generalized conjugacy problem* GCP(m) for G asks for an algorithm to decide for any sequence of length m whether or not the sequence is dependent. Note that GCP(1) is the word problem and GCP(2) is the "ordinary" conjugacy problem.

We remark that a slightly different family of decision problems (the "dependence problems") was introduced in [23] (see also [8], [25]). The connection with geometry is as follows.

Consider objects D which can be built up by means of an ascending union.

$$D_0 \subseteq D_1 \subseteq \cdots \subseteq D_n = D,$$

where D_0 is a single vertex, and where D_{i+1} is obtained from D_i either by attaching a *spine* (that is, a finite combinatorial subdivision of a closed interval) to D_i by one of its endpoints to a vertex of D_i, or by attaching a *finitely tesselated sphere* by one of its vertices to a vertex of D_i. A *labelling* of such an object D is an assignment of an element of $\mathbf{x} \cup \mathbf{x}^{-1}$ to each oriented edge of D, with the understanding that if an oriented edge is assigned label $t \in \mathbf{x} \cup \mathbf{x}^{-1}$ then the oppositely oriented edge is assigned label t^{-1}. We call D together with a labelling a *P-diagram* (or diagram for short).

For each region Δ of a diagram we fix a vertex (basepoint) on $\partial\Delta$. The word we read off by travelling once around $\partial\Delta$ in the anticlockwise direction starting at the basepoint is called the *label* on Δ. We say that Δ is *distinguished* if its label is not a cyclic permutation of an element of $\mathbf{r} \cup \mathbf{r}^{-1}$.

A diagram is *reduced* if whenever Δ, Δ' are two non-distinguished regions meeting along a common edge e, the word we read off by travelling once around $\partial\Delta$ anticlockwise starting with e is not the same as the word we read off by travelling once around $\partial\Delta'$ clockwise starting with e.

Theorem. $(W_1, W_2, \cdots, W_k)$ *is dependent if and only if there is a reduced diagram with exactly k distinguished regions, where the labels on the distinguished regions are* $W_1, W_2, \cdots, W_k$.

This can be proved by arguments similar to those in §§V.1, V.5 of [16].

Problem 6. Investigate the generalized conjugacy problems for groups given by various types of presentations.

For example, by slightly adapting arguments given in [23], one can show that GCP(m) is solvable for all m, for groups given by finite hyperbolic presentations, where "hyperbolic" is defined as follows.

The *star-complex* P^{st} of P is the 1-complex with vertex-set $\mathbf{x} \cup \mathbf{x}^{-1}$ and edge-set the set $\mathbf{r}^*$ of all cyclic permutations of elements of $\mathbf{r} \cup \mathbf{r}^{-1}$. For an edge $R \in \mathbf{r}^*$, the initial vertex is the first symbol of R, and the terminal vertex is the inverse of the last symbol of R. The edge inverse to R is just R^{-1}. A *weight function* on P^{st} is a function θ from the edge-set $\mathbf{r}^*$ to the set of non-negative real numbers, such that $\theta(R^{-1}) = \theta(R)$ for all $R \in \mathbf{r}^*$. The *weight of a path* is then the sum of the weights of the edges making up the path. We say that P is *hyperbolic* if there is a weight function θ on P^{st} such that:

(i) the weight of every non-empty cyclically reduced closed path is at least 2;

(ii) for each element $y_1 y_2 \cdots y_n \in \mathbf{r}$ (where $y_1, y_2, \cdots, y_n \in \mathbf{x} \cup \mathbf{x}^{-1}$),

$$\sum_{i=1}^{n}(1 - \theta(y_i \cdots y_n y_1 y_2 \cdots y_{i-1})) > 2.$$

We note that the above concept of hyperbolicity is due to Gersten [10] (see also [23]). More general notions of hyperbolicity are considered by Gromov in [11].

Problem 7. Is there a (finitely generated, presented) group for which GCP(m) is solvable but GCP($m + 1$) is not solvable?

The answer is, of course, well-known to be "yes" when $m = 1$ (see [18] for a discussion of this).

4. Asphericity.

A P-diagram with no distinguished regions is usually called a *spherical* P-diagram. Such diagrams are of importance in connection with various notions of asphericity in group theory. The basic reference for a discussion of asphericity is [5] (see also [3], [6]). We will be interested here in two types of asphericity.

Diagrammatic asphericity. There are certain operations which one can perform on spherical P-diagrams called *diamond moves* – see [6, §1.4] for the definition. A presentation P is said to be *diagrammatically aspherical* (DA) if every spherical P–diagram having at least one sphere can be converted by a finite sequence of diamond moves to a spherical P-diagram in which one of the spheres has just two regions.

Cohen-Lyndon asphericity. We say that $P = \langle \mathbf{x} \ ; \ \mathbf{r} \rangle$ is *Cohen-Lyndon aspherical* (CLA) if the normal closure N of $\mathbf{r}$ in the free group F on $\mathbf{x}$ has a free generating set

$$\{WRW^{-1} : R \in \mathbf{r}, \ W \text{ belongs to a transversal for } NC_F(R) \text{ in } F\}.$$

Here $C_F(-)$ is the centralizer in F.

If G is the group defined by P then we often abuse terminology and say that G is DA or CLA if P is.

Now it can be proved that CLA $\Longrightarrow$ DA (see [5, Proposition 1.7]). However, whether or not the reverse implication is true remains unresolved. (This is the only implication between the various notions of asphericity discussed in [5] which remains unresolved.)

A method of constructing DA groups is given in [24]. Let Γ be a graph with vertex set $\mathbf{x}$ and edge set e. For each edge $\{x, y\} \in$ e let $\mathbf{r}\{x, y\}$ be a set of cyclically reduced words on x and y (each word involving both x and y). Let

$$G\{x, y\} = \langle x, y ; \mathbf{r}\{x, y\}\rangle.$$

We say that $G\{x, y\}$ has *property-W_p* if no word of the form $x^{\alpha_1} y^{\beta_1} \cdots x^{\alpha_p} y^{\beta_p}$ ($\alpha_1, \beta_1, \cdots \alpha_p, \beta_p \in \mathbf{Z}$) is equal to 1 in $G\{x, y\}$ unless it is freely equal to 1.

Theorem [24]. *Assume that one of the following conditions is satisfied.*

(I) $G\{x, y\}$ has property-W_2 for each $\{x, y\} \in$ e.

(II) Γ is triangle-free and each $G\{x, y\}$ ($\{x, y\} \in$ e) has property-W_1.

Then the group

$$G = \langle \mathbf{x} ; \bigcup \{\mathbf{r}\{x, y\} \mid \{x, y\} \in \mathbf{e}\}\rangle$$

is DA if and only if each of the groups $G\{x, y\}$ ($\{x, y\} \in$ e) is DA.

Problem 8. Can one use the above construction to give examples of groups which are DA but not CLA?

Of course, the above construction would only be of use provided the Cohen-Lyndon asphericity of the groups $G\{x, y\}$ ($\{x, y\} \in$ e) did not in general imply the Cohen-Lyndon asphericity of G. Therefore tackling Problem 8 really amounts to tackling the problem of whether or not a theorem similar to the above holds, with DA replaced by CLA throughout.

An interesting special case of Problem 8 concerns *Artin groups*. Consider a pair (Γ, ϕ) as in §1. The Artin group $A(\Gamma, \phi)$ associated with this pair is the group given by the presentation

$$\langle \mathbf{x} ; \underbrace{x \, y \, x \cdots \cdots}_{\phi\{x,y\} \text{ terms}} = \underbrace{y \, x \, y \cdots \cdots}_{\phi\{x,y\} \text{ terms}} \quad (\{x, y\} \in \mathbf{e})\rangle$$

Condition (I) holds if the Artin group is of *large type* [1], [2] (that is, if $2 \notin \operatorname{Im} \phi$, while condition (II) holds if the Artin group is of *triangle-free type* [21] (that is, if Γ is triangle-free).

Problem 8'. Are Artin groups of large type and triangle-free type CLA?

5. Finite groups with balanced presentation.

Let G be a group given by a presentation with m generators and n defining relators. If $m > n$ then G is infinite (in fact, G has an infinite cyclic homomorphic image). On the other hand, if $m = n$ (that is, if the presentation is *balanced*) then we have a sort of borderline situation in that G may be finite or infinite. The question of exactly when a balanced presentation defines a finite group, and exactly which finite groups can arise, is far from being settled.

All known finite groups with balanced presentations have rank at most 3 (see [12], [13]). (The *rank* of a group is the least number of generators.)

Problem 9. (Well-known problem) Is there a finite group of rank greater than 3 having a balanced presentation?

It should be noted that for any finite group G having a balanced presentation the *abelianization* G/G' of G has rank at most 3 [14, Theorem 9 (ii)].

Many of the examples of balanced presentations defining finite groups of rank 3 have the form

$$\langle x_1, x_2, x_3 \; ; \; R_1(x_1, x_2), R_2(x_2, x_3), R_3(x_3, x_1) \rangle.$$

(See [9], [13, §3], [15].). In view of this, the following problem seems reasonable.

Problem 9'. Let

$$G = \langle x_1, x_2, \cdots x_n \; ; \; R_1(x_1, x_2), R_2(x_2, x_3), \cdots, R_n(x_n, x_1) \rangle$$

with $n \geq 4$. Find the rank of G when G is finite.

This problem has been solved [22, Theorem 3] in the case when the presentation of G is *cyclic* (that is, when there is a word $R(a, b)$ in two variables a, b such that $R_i(x_i, x_{i+1}) = R(x_i, x_{i+1})$ for $i = 1, 2, \cdots, n$ (computed mod n). In this case, if G is finite, then G is cyclic.

We note that Problem 9 above is related to Problem B4 of [29].

6. One relator groups with coefficients

Let A be a group and let $\mathbf{x}$ be a set. Let R be a "word" on $\mathbf{x}$ and A (that is, R is an element of the free product of A and the free group on $\mathbf{x}$). The $\mathbf{x}$-*skeleton* of R is the word obtained from R by removing all the terms belonging to A. For example, if $R = a_1 x^2 y a_2 y^{-1} x^{-1}$ with a_1, $a_2 \in A$ and x, $y \in \mathbf{x}$, then the x-skeleton is $x^2 y y^{-1} x^{-1}$. We have the *one-relator group with coefficients*

$$G = \langle A, \mathbf{x} \; ; \; R \rangle.$$

We will say that the *Freiheitssatz holds* for G if for each subset $\mathbf{x}_0$ of $\mathbf{x}$ omitting a generator occurring in R, the natural homomorphism of the free product $A * F_0$ (where F_0 is the free group on $\mathbf{x}_0$) into G is injective.

Theorem. *The Freiheitssatz holds for G if the x-skeleton of R is non-empty and cyclically reduced.*

The proof of this is modelled on McCool and Schupp's proof [17] for ordinary one-relator groups. The main case of the induction step is when some element of $\mathbf{x}$ occurs in R with zero exponent sum. In this case one can write G as an HNN-extension of an one-relator group $H = \langle B, \mathbf{x}' \; ; \; P \rangle$ with coefficients, where the coefficient group B is a free product of copies of A, and the $\mathbf{x}'$-skeleton of P has length less than the $\mathbf{x}$-skeleton of R. An example should suffice to show what is involved here. Let

$$G = \langle A, t, x, y \; ; \; a_1 x t^3 a_2 x^2 y^{-2} a_3 t^{-1} x a_4 y t^{-2} \rangle,$$

where $a_1, a_2, a_3, a_4 \in A$. For $i \in \mathbf{Z}$ let $A^{(i)}$ be a copy of A (where $a^{(i)} \in A^{(i)}$ corresponds to $a \in A$), and let

$$B = \mathop{*}_{i \in \mathbf{Z}} A^{(i)}.$$

Let

$$H = \langle B, x_0, x_1, x_2, x_3, y_i \; (i \in \mathbf{Z}) \; ; \; a_1^{(0)} x_0 a_2^{(3)} x_3^2 y_3^{-2} a_3^{(3)} x_2 a_4^{(2)} y_2 \rangle.$$

Then G is an HNN-extension of H with stable letter t and associated subgroups

$$sgp\{B, x_0, x_1, x_2, y_i \; (i \in \mathbf{Z})\}, \quad sgp\{B, x_1, x_2, x_3, y_i \; (i \in \mathbf{Z})\}.$$

For ordinary one-relator groups the *initial* step of the inductive proof of the Freiheitssatz is completely trivial. However, for one-relator groups with coefficients this is no longer the case in general. The initial step is the assertion that if $G = \langle A, t \; ; \; R \rangle$ where t occurs in R and only to positive powers, then the natural homomorphism $A \longrightarrow G$ is injective. This is a theorem of F. Levin [15].

Now an obvious question is whether one can solve the word problem for one-relator groups with coefficients when the $\mathbf{x}$-skeleton is cyclically reduced and non-empty (assuming that the coefficient group has solvable word problem). If one wants to mimick the proof for ordinary one-relator groups, then, in fact, one needs to consider a more general problem, the analogue of the so-called *Magnus problem* for one-relator groups.

Let the coefficient group A be given by a presentation $\langle \mathbf{y}; \mathbf{s} \rangle$ (with solvable word problem), so that

$$G = \langle \mathbf{x}, \mathbf{y} \; ; \; R, \mathbf{s} \rangle.$$

(We assume that $\mathbf{x}$ and $\mathbf{y}$ are finite, or countably infinite). We ask for an algorithm to decide for any word W on the generators of G, and for any recursive subset $\mathbf{x}_0$ of $\mathbf{x}$ omitting a generator occurring in R, whether or not W defines an element of sgp $\mathbf{x}_0 \cup \mathbf{y}$, and if so to give an expression for this element in terms of $\mathbf{x}_0 \cup \mathbf{y}$.

Again, one can push the induction step through, but again it is the initial step which is a problem. Thus we have:

Problem 10. Let $G = \langle t, \mathbf{y} \; ; \; R, \mathbf{s} \rangle$ where R involves t, and all powers of t are positive. Assume that the word problem for $A = \langle \mathbf{y} \; ; \; \mathbf{s} \rangle$ is solvable. Is there an algorithm to decide for any word W on the generators of G, whether or not W is equal in G to a word on $\mathbf{y}$?

References

[1] K.I. Appel, 'On Artin and Coxeter groups of large type', *Contemporary Mathematics 33* (Amer. Math. Soc., Providence, 1984), 50-78.

[2] K.I.Appel and P.E. Schupp, 'Artin groups and infinite Coxeter groups', *Invent. Math.*, 72(1983), 201-220.

[3] R. Brown and J. Huebschmann, 'Identities among relations' in: *Low Dimensional Topology*, edited by R. Brown and T.L. Thickstun, LMS Lecture Note Series 48, CUP, 1982.

[4] N. Bourbaki, 'Groupes et Algèbres de Lie, Chapitre 4, 5 et 6', *Actu. Sci. Ind.* n° 1337, Hermann, Paris, 1968.

[5] I.M. Chiswell, D.J. Collins and J. Huebschmann, 'Aspherical group presentations', *Math. Z.*, 178(1981), 1-36.

[6] D.J. Collins and J. Huebschmann, 'Spherical diagrams and identities among relations', *Math. Ann.*, 261(1982), 155-183.

[7] M.W. Davis, 'Groups generated by reflections and aspherical manifolds not covered by Euclidean space', *Ann. Math.*, 117(1983), 293-324.

[8] M.S. El-Mosalamy and S.J. Pride, 'On T(6)-groups', *Math. Proc. Camb. Phil. Soc.*, 102(1987) 443-451.

[9] W. Fluch, 'A generalized Higman group', *Nederl. Acad. Wetensch. Indag. Math.*, 44(1982), 153-156.

[10] S.M. Gersten, 'Reducible diagrams and equations over groups' in: *Essays in Group Theory*, edited by S.M. Gersten, Springer-Verlag, 1987.

[11] M. Gromov, 'Hyperbolic groups' in: *Essays in Group Theory*, edited by S.M. Gersten, Springer-Verlag, 1987.

[12] D.L. Johnson, *Topics in the Theory of Group Presentations*, Cambridge University Press, Cambridge, 1980.

[13] D.L. Johnson and E.F. Robertson, 'Finite groups of deficiency zero' in: *Homological Group Theory*, edited by C.T.C. Wall, Cambridge University Press, Cambridge, 1979.

[14] D.L.Johnson, J.W. Wamsley and D. Wright, 'The Fibonacci groups', *Proc. London Math. Soc.*, 20(1974), 577-592.

[15] F.S. Levin, 'Solutions of equations over groups', *Bull. Amer. Math. Soc.*, 68(1962), 603-604.

[16] R.C. Lyndon and P.E. Schupp, *Combinatorial Group Theory*, Springer-Verlag, Berlin, 1977.

[17] J. McCool and P.E. Schupp, 'On one-relator groups and HNN-extensions', *J. Austral. Math. Soc.*, 16(1973), 249-256.

[18] S. Meskin, 'Non-residually finite one-relator groups', *Trans. Amer. Math. Soc.*, 164(1972), 105-114.

[19] C.F. Miller III, *On Group-Theoretic Decision Problems and their Classification*. Princeton University Press, Princeton, 1971.

[20] M.J. Post, 'Finite three-generator groups with zero deficiency', *Comm. Algebra*, 6(1978), 1289-1296.

[21] S.J. Pride, 'On Tits' conjecture and other questions concerning Artin and generalized Artin groups', *Invent. Math.*, 86(1986), 347-356.

[22] S.J. Pride, 'Groups with presentations in which each defining relator involves exactly two generators', *J. London Math. Soc.* (2), 36(1987), 245-256.

[23] S.J. Pride, 'Star-complexes, and the dependence problems for hyperbolic complexes', *Glasgow Math. J.*, 30(1988), 155-170.

[24] S.J. Pride, 'The diagrammatic asphericity of groups given by presentations in which each defining relator involves exactly two types of generators', *Arch. Math.*, 50(1988), 570-574.

[25] S.J. Pride, 'Involutary presentations, with applications to Coxeter groups, NEC-groups and groups of Kanevskiï', *J. Algebra*, to appear.

[26] S.J. Pride and R. Stöhr, 'The (co)homology of aspherical Coxeter groups', preprint.

[27] P.E. Schupp, 'A survey of small cancellation theory' in: *Word Problems*, North-Holland, 1973.

[28] M. Suzuki, *Group Theory I*, Springer-Verlag, Berlin, 1982.

[29] C.T.C. Wall (ed), *Homological Group Theory*, Cambridge University Press, Cambridge, 1979.

Department of Mathematics
University of Glasgaw
University Gardens
G12 8SQ
Scotland

GROUPS WITH MANY ELLIPTIC SUBGROUPS

Akbar H. Rhemtulla

Abstract. We say that a group G has many elliptic pairs of subgroups if every infinite set $\{H_1, H_2, \cdots\}$ of subgroups of G contains at least one pair H_i, H_j, $i \neq j$ such that $\langle H_i, H_j \rangle = (H_i H_j)^n$ for some integer n depending on H_i, H_j. By $(H_i H_j)^n$ we mean the set of all products of $2n$ elements each from H_i or H_j. Let G be a finitely generated solvable group. It is shown that G has many elliptic pairs of subgroups if and only if G is finite-by-nilpotent. It is also shown that if G is finitely generated, torsion-free and residually finite p-group, for some prime p, then G has many elliptic pairs of subgroups if and only if G is nilpotent.

1. Introduction.

1.1. A subgroup H of a group G is called *elliptic* if for each subgroup K of G there is an integer n (depending on K) such that

$$\langle H, K \rangle = HK \cdots HK$$

where the product has $2n$ factors. A normal subgroup is elliptic and so is a quasinormal subgroup since $H \leq G$ is quasinormal if $HK = KH$ for all $K \leq G$. Our use of the word elliptic is motivated by the terminology due to Philip Hall given in his lectures in Cambridge in the 1960's. He called a set X an elliptic set if there is an integer n such that every element of the group generated by X can be written as a product of at most n elements of $X \cup X^{-1}$.

1.2. If G is a solvable group generated by the set $\{g_1, \cdots, g_n\}$ then it is finite-by-nilpotent if and only if each $\langle g_i \rangle$ is elliptic in G (Theorem 1, [9]). From this and using recent results of Lubotzky and Mann it can be shown that if G is a finitely generated residually finite p-group for some prime p, then G is finite-by-nilpotent if every subgroup of G is elliptic.

1.3. Let G be a group and $\mathcal{F}$ the family of all subgroups of G. If $H, K \in \mathcal{F}$ and $\langle H, K \rangle = (HK)^n$ for some n, then we call H, K an *elliptic pair*. A group G has many elliptic pairs if every infinite subset of $\mathcal{F}$ contains an elliptic pair. Groups with many pairs with prescribed property have been studied before. B.H. Neumann [8] proved that if every infinite subset of elements of G has a commuting pair of elements then G is center-by-finite. Other problems of this type are contained in [5] and [3]. In a recent paper [1] it was shown that if every infinite set of subgroups of a finitely generated solvable group G has a permuting pair of subgroups then G is center-by-finite. A corollary of this is that a torsion-free group with many permuting pairs of subgroups is abelian. We can look at groups with many elliptic pairs as a more general class of groups compared to above.

1.4. The main results of this paper are as follows.

Theorem A. *Let G be a finitely generated solvable group. Then G has many elliptic pairs of subgroups if and only if G is finite-by-nilpotent.*

Theorem B. *Let G be finitely generated, torsion-free, residually finite p-group for some prime p. Then G has many elliptic pairs of subgroups if and only if G is nilpotent.*

Theorems A and B have the following corollaries.

Corollary C. *Let G be a finitely generated solvable group. If G has many elliptic pairs of subgroups then every pair of subgroups is elliptic.*

Corollary D. *Let G be a finitely generated torsion-free residually finite p-group for some prime p. If G has many elliptic pairs of subgroups then every pair of subgroups is elliptic.*

We have not considered the period case, but present the following:

Conjecture. A periodic group with many elliptic pairs of subgroups is locally finite.

2. Preliminary results and proof of Theorem A.

2.1. Note that If G has infinitely many elliptic pairs of subgroups then so does any quotient and subgroup of G. The following results will be needed in the proof of Theorem A.

Lemma 2.2. *Let $G = \langle a \rangle$ wreath $\langle t \rangle$, the wreath product of a cyclic group $\langle a \rangle$ of prime order p and an infinite cyclic group $\langle t \rangle$. Then G does not have many elliptic pairs of subgroups.*

Proof. The base group B of G is isomorphic to the additive group of the group algebra $\mathbf{F}_p\langle t \rangle$. Let $h_i = t^{a t^i}$, $i = 0, 1, \cdots$ and $H_i = \langle h_i \rangle$. If G were to have many elliptic pairs of subgroups then for some $0 \le i < j$, $\langle H_i, H_j \rangle = (H_i H_j)^r$ for some integer r. But this implies $\langle H_0, H_k \rangle = (H_0 H_k)^r$ where $k = j - i > 0$. Observe that every element of $(H_0 H_k)^r \cap B$ has the form $\sum_{i=0}^{2r}(1 - t^{\alpha_i})t^{\lambda_i}$ for suitable integers α_i, λ_i. These are sums of bounded number of terms. On the other hand the subgroup $< H_0, H_k >^r \cap B$ is of finite index in B and hence contains sums of above form of arbitrarily large number of terms. Thus G can not have many elliptic pairs.

Lemma 2.3. *Let G be a finitely generated nilpotent-by-finite solvable group with many elliptic pairs of subgroups. Then G is finite-by-nilpotent.*

Proof. By hypothesis G contains a normal nilpotent subgroup N such that G/N is finite. Take N to be maximal subject to being nilpotent and normal in G. Use induction on the order $|G/N|$ of G/N. If $N = G$ then there is nothing to prove. Since G/N is solvable, there is a subgroup $M \lhd G$, $N \le M$ and $|G/M| = p$, a prime. By induction hypothesis, M is finite-by-nilpotent. It is also finitely generated since G is finitely generated. Hence the periodic elements of M form a finite normal subgroup T of G. Since we want to show that G is finite-by-nilpotent, we may assume $T = 1$ by considering G/T in place of G if necessary. Thus $G = \langle M, t \rangle$ where M is a torsion-free nilpotent normal subgroup of G and $t^p \in M$.

Let $A \leq B$ be normal subgroups of G such that B/A is a central factor of M. Let $\bar{G} = G/A$ and consider the subgroup $\langle B, t \rangle/A$ of $\bar{G}$. If $\bar{t}$ has only finitely many conjugates in $\langle B, t \rangle/A$, then the centralizer $\bar{C}$ of $\bar{t}$ in $\langle B, t \rangle/A$ is of finite index in $\langle B, t \rangle/A$. Its intersection with $\langle B, t^p \rangle/A$ is of finite index in $\langle B, t \rangle/A$ and lies in its center. Thus, by Schur's Lemma, $\langle B, t \rangle/A$ is finite-by-abelian. Now

$$[G, B] = [M \langle t \rangle, B] = [M, B][\langle t \rangle, B] \leq A \langle B, t \rangle'$$

Thus $A[G, B]/A$ is finite. In particular if At has only finitely many conjugates in $\langle B, t \rangle/A$ for every central factor B/A of M where A, B are normal in G, then $\gamma_c(G)$ is finite if $\gamma_c(M) = 1$.

Thus assume that for some central factor B/A of M, At has infinitely many conjugates in $\langle B, t \rangle/A$. We may as well assume that B/A is torsion-free since we can always replace A by T where T/A is the torsion subgroup of B/A. Now identify A with 1 so the case to be considered is as follows.

$\langle B, t^p \rangle$ is abelian and t has infinitely may conjugates in $\langle B, t \rangle$. Since B is finitely generated, there is $b \in B$ such that $\langle t \rangle, \langle t^b \rangle, \langle t^{b^2} \rangle, \cdots$ are all distinct. By hypothesis, this set of subgroups contains an elliptic pair. Hence for some integer $k > 0$, $\langle t, t^{b^k} \rangle = (\langle t \rangle \langle t^{b^k} \rangle)^n$ for some n. Replace b by b^k, if necessary, and assume $\langle t, t^b \rangle = (\langle t \rangle \langle t^b \rangle)^n$. Now $(\langle t \rangle \langle t^b \rangle)^n$ lies in a finite number of cosets of $\langle t^p \rangle$. Thus $\langle t, t^b \rangle$ is a finite extension of $\langle t^p \rangle$, hence $(t^{-1} t^p)^m \in \langle t^p \rangle$. Since b commutes with all its conjugates, $[t, b^m] = [t, b]^m$ and so $t^{b^m} \in \langle t \rangle$, hence $\langle t \rangle = \langle t^{b^m} \rangle$, a contradiction. This completes the proof.

Proof of Theorem A. If G is finitely generated finite-by-nilpotent then every pair of subgroups of G is elliptic (Proposition 1, [9]). It is the converse that we need to show. Let G be a finitely generated solvable group with many elliptic pairs of subgroups. Use induction on the solvability length of G to show that G is finite-by-nilpotent. If G has solvability length one then it is abelian and there is nothing to prove. Assume the result for groups of solvability length at most d and suppose $G^{(d+1)} = 1$ where $G^{(r)}$ denote the rth term of the derived series of G. Then $G/G^{(d)}$ is finite-by-nilpotent. Since G is finitely generated, there exists a normal subgroup N of finite index in G such that $N \geq G^{(d)}$ and $N/G^{(d)}$ is nilpotent.

If we show that N is finite-by-nilpotent then it will follow from Lemma 2.3 that G is finite-by-nilpotent as required. Thus assume that G is abelian-by-nilpotent. By an easier special case of Kropholler's main result in [4], and Lemma 2.2 it follows that G has finite (Prüfer) rank. By a result of P. Hall, G satisfies max–n since it is a finitely generated abelian-by-nilpotent group. Thus any normal torsion subgroup of G is finite.

Now G has a normal subgroup B such that G/B is torsion-free nilpotent and B is abelian-by-finite. It follows from Lemma 2.3 that the set T of torsion elements of B form a subgroup. By what was stated in the above paragraph, T is finite. Thus we may take T to be the identity subgroup since we want to show that G is finite-by-nilpotent. Hence assume that G is torsion-free, has a normal abelian subgroup A and G/A is torsion-free nilpotent. We show that G is nilpotent.

If $\langle A, t \rangle$ is nilpotent for every $t \in G$, then so is G. Thus we may assume that $G = \langle A, t \rangle$. Let $A_1 \neq 1$ be an isolated subgroup of A of smallest rank such that t^n normalizes A_1 for some integer $n \neq 0$. Suppose that $[t^n, A_1] = 1$. Then $\langle A_1, t \rangle$ is abelian-by-finite and torsion-free. Hence by Lemma 2.3, it is nilpotent and hence abelian. Thus if we show that $\langle A_1, t^n \rangle$ is abelian then take the quotient G/A_1 and repeat the process, we would, after a finite number of steps, get A to be in the hypercentre of G, and hence G to be nilpotent as required.

Thus we may assume that $G = \langle A, t \rangle$ and every nontrivial subgroup of A normalized by some power t^n, $n \neq 0$, is of finite index in A. Thus t acts rationally irreducibly on A. Pick any $a \neq 1$ in A. Then $\langle a, t \rangle$ is of finite index in G. Consider the set $\langle t \rangle$, $\langle t^a \rangle$, $\langle t^{a^2} \rangle, \cdots$ of subgroups of G. If this set is finite then $t^{a^n} = t$ for some n and by what we have said above, G is abelian as required. Thus suppose that this set is infinite. Then it follows from the hypothesis that $\langle t, t^{a^n} \rangle = (\langle t \rangle \langle t^{a^n} \rangle)^k$ for some integer k. Put $b = a^n$. Then every element of $\langle t, t^b \rangle$ is a product of k terms of the form $t^{\lambda + \mu} b b^{-t^\lambda}$. Thus each element of $A \cap \langle t, t^b \rangle$ is a product of $2k$ conjugates of $b^{\pm 1}$ under $\langle t \rangle$.

Using additive notation for A, each element of $A \cap \langle t, t^b \rangle$ has the form

$$b \left(\sum_{i=1}^{2k} \pm t^{\lambda_i} \right)$$

with $\lambda_i \in \mathbf{Z}$. Now $V = A \otimes_{\mathbf{Z}} \mathbf{Q}$ is an irreducible $\mathbf{Q}\langle t \rangle$ module, and by Schur's Lemma the centralizer ring $\Gamma = \mathrm{End}_{\mathbf{Q}\langle t \rangle} V$ is a division ring, finite dimensional over $\mathbf{Q}$. The image of t in $\mathrm{End}_{\mathbf{Q}} V$ lies in and spans Γ so that Γ is an algebraic number field. Further, regarded as a Γ-vector space, V is of dimension one. Let τ be the image of t in Γ, and choose $d \neq 0$ from $A \cap \langle t, t^b \rangle$ so that $d = b\phi$ for some $\phi \in \Gamma$. Thus for each integer m we can write $mb\phi$ in the form

$$b \sum_{i=1}^{2k} \pm \tau^{\lambda_i}$$

so that $m\phi$ has the form

$$\sum_{i=1}^{2k} \pm \tau^{\lambda_i}.$$

Now τ is not a root of 1 since $[t^n, A] \neq 1$ and t^n acts rationally irreducibly on A for all $n \neq 0$. Thus Γ can be embedded in $\mathbf{C}$ so that $|\tau| > 1$. It follows from Lemma 2 of [9] or more generally from a theorem of Evertse [2] or Van der Poorten [11] that not every $m\phi$ can be expressed as a bounded sum of elements of the form $\pm \tau^{\lambda_i}$. This completes the proof.

3. Preliminary results and proof of Theorem B.

Lubotzky and Mann have recently discussed the linearity of groups which are almost residually finite p-groups for some prime p. These beautiful results have made the theorem in this section possible.

Lemma 3.1. Let H, K be finitely generated nilpotent groups. If $G = \langle H, K \rangle = (HK)^n$ for some integer n and G is residually finite p-group for some prime p, then G is a linear group over $\mathbf{C}$.

Proof. For each positive integer m, $|G/G^{p^m}| \leq p^{lm}$ for some fixed l depending on n, the nilpotency classes of H and K and number of generators of H and K. This may be seen as follows. By Proposition 2(a) of [9], for all subgroups A, B of H, $\langle A, B \rangle = (AB)^t$ where $t = (4r)^c$ and r, c are respectively the rank and class of the nilpotent group H. Thus if $H = \langle h_1, \cdots h_r \rangle$ and we write $H_i = \langle h_i \rangle$, then $\langle h_1, h_2 \rangle = (H_1 H_2)^t$, and more generally $\langle h_1, \cdots, h_{i+1} \rangle = (H_1 \cdots H_{i+1})^{t^i}$ for all $i \geq 1$. Thus $H = (H_1 \cdots H_r)^\lambda$ where $\lambda = t^{r-1}$. Similarly if $K = \langle K_1, \cdots K_s \rangle$ and $K_i = \langle K_1 \rangle$ then $K = (K_1 \cdots K_s)^\mu$ for a suitable integer μ. Now $G = (HK)^n = (H_1 \cdots H_r K_1 \cdots H_s)^l$ where $l = \lambda \mu n$.

Since $G^{p^m} \geq \langle h_1^{p^m}, \cdots, k_s^{p^m} \rangle$ and $|H_i/H_i^{p^m}| \leq p^m$, $|G/G^{p^m}| \leq p^{lm}$. Now pick any positive integer $c > 4l$ and let $J = G/G^{p^c}$. Let $\Phi_1(J)$ denote the Frattini subgroup of J and for $i > 1$ let $\Phi_i(J) = \Phi_1(\Phi_{i-1}(J))$. Then for some $i \leq 4l$, $d(\Phi_i(J)) < i$ where $d(G)$ denotes the minimal number of generators of a group G. If this were not so then $d(\Phi_i(J)) \geq i$ for each $i = 1, 2, \cdots, 4l$ and hence $|J/\Phi_{4l}(J)| \geq p^{1+2+\cdots+4l} > p^{8l^2}$. But $\Phi_i(J) \geq J^{p^i}$ for all $i \geq 1$. Hence

$$p^{4l^2} \geq |J/J^{p^{4l}}| \geq |J/\Phi_{4l}(J)| > p^{8l^2},$$

a contradiction. In the terminology of Lubotzky and Mann, $\Phi_{4l}(J)$ is a powerful p-group (1.2, [7]) and it follows from Lazard's Theorem (see Theorem 2.1 of [7]) or also from the main result of Lubotzky [6], that $G \subseteq GL(\mathbb{C})$.

Lemma 3.2. *Let H, K be finitely generated nilpotent groups, $G = \langle H, K \rangle = (HK)^n$ for some integer n and G is residually finite p-group with many elliptic pairs of subgroups. Then G is finite-by-nilpotent.*

Proof. By Lemma 3.1, G is a linear group. It can not have a non-abelian free group as a subgroup since such a group does not have many elliptic pairs of subgroups. Thus by Tits result (Theorem 1, [10]) it is solvable by finite. As G is also residually finite p-group, it is solvable. Theorem A now applies and we get the result.

Lemma 3.3. *Let G be a torsion-free residually finite p-group, p a prime, with many elliptic pairs of subgroups. If $x \in H \leq G$ and $x^n \in Z(H)$ for some $n \neq 0$, then $x \in Z(H)$.*

Proof. Note that $C_H(x) = N_H(\langle x \rangle)$ for otherwise G would have a section isomorphic to the infinite dihedral group which has infinitely many subgroups of order two with no elliptic pair.

Let $C = C_H(x)$ and let T be a set of coset representatives of C in H. Suppose T is infinite. Form a sequence of conjugates $x, x^{t_2}, x^{t_3}, \cdots$ of x as follows: the first member of the sequence is x. If there is $t_2 \in T$ such that $\langle x \rangle$, $\langle x^{t_2} \rangle$ is not an elliptic pair, then put x^{t_2} as the next member. If $x, x^{t_2}, \cdots, x^{t_m}$ have been chosen and there is $t_{m+1} \in T$ such that $\langle x^{t_i} \rangle$, $\langle x^{t_{m+1}} \rangle$ is not an elliptic pair for all $i \leq m$ ($x^{t_1} = x$), then put $x^{t_{m+1}}$ as the next element of the sequence. This process must terminate after a finite number, say d, of steps since G has many elliptic pairs.

For each $i \leq d$ let T_i be the set of those $t \in T$ such that $\langle x^{t_i} \rangle$, $\langle x^t \rangle$ form an elliptic pair. Then $T = T_1 \cup T_2 \cup \cdots \cup T_d$. By Lemma 3.2, $t \in T_i$ implies $\langle x^t, x^{t_i} \rangle$ is nilpotent and since $x^n \in Z(H)$, $x^t = x^{t_i}$. Thus $t = t_i$ since T is a transversal to C in H. Thus T is finite and x has finitely many conjugates in H.

Now $\langle x^H \rangle$ is a torsion-free FC group and therefore abelian. It has rank one since $x^n \in Z(H)$. Thus $\langle x^H \rangle = \langle x \rangle$ as required.

Proof of Theorem B. If G is finitely generated nilpotent then every pair of subgroups of G is elliptic, as remarked in the proof of Theorem A. Implication the other way is non-trivial and we prove this. By hypothesis $G = \langle x_1, \cdots, x_r \rangle$ is a torsion-free residually finite p-group for some prime p and G has many elliptic pairs of subgroups. If $r = 1$ then G is cyclic, hence nilpotent as required. Consider next the case where $r = 2$. Then $G = \langle x, y \rangle$. Let $H_i = \langle (x^i y) \rangle$, $i = 0, 1, \cdots$. If $H_i = H_j$ for some $i < j$, then $\langle x^i y, x^j y \rangle = \langle x^i y, x^{j-i} \rangle$ is abelian and hence x^{j-i} commutes with y. By Lemma 3.3, x lies in the centre of $\langle x, y \rangle$ and $\langle x, y \rangle$ is abelian as required.

We may now suppose $H_i \neq H_j$ if $i \neq j$. Then for some $i < j$, H_i, H_j is an elliptic pair, and by Lemma 3.2, $\langle H_i, H_j \rangle = \langle x^i y, x^k \rangle$ is nilpotent where $k = j - i > 0$. So there exists $1 \neq z$ in the center of $\langle x^i y, x^k \rangle$. Now consider $\langle x, z \rangle$. Since x^k lies in the centre of this group, by Lemma 3.3, x lies in the center of $\langle x, z \rangle$. Hence $\langle x, z \rangle$ is abelian and z lies in the center of $G = \langle x, y \rangle$. In particular $Z(G) \neq 1$ if $G \neq 1$. Again by Lemma 3.3, $Z(G)$ is an isolated subgroup of G so that $G/Z(G)$ is again torsion-free. Now the property of being residually finite p-group is preserved when we take the quotient $G/Z(G)$. Thus all the hypotheses of the theorem hold for $G/Z(G)$ and we conclude that G is a hypercentral group. But G is finitely generated. Hence G is nilpotent, as required, and we have dealt with the case $r = 2$.

Now assume, by way of induction, that we have dealt with the case where the group is generated by fewer than r elements, and $G = \langle x_1, \cdots, x_r \rangle$. For each $i \geq 0$ let $J_i = \langle x_1, \cdots, x_{r-2}, (x_{r-1} x_r^i) \rangle$. If $J_i = J_k$ for some $k \neq i$, then $x_r^{k-i} \in J_i$ which is nilpotent by induction hypothesis. The centre Z of J_i commutes with x_r^{k-i} and, by Lemma 3.3, it therefore commutes with x_r and hence with x_{r-1}. Thus Z lies in the center of G. By Lemma 3.3, $Z(G)$ is isolated thus G is nilpotent of the same class as that of J_i and we are done.

We may now suppose that $J_i \neq J_k$ if $i \neq k$. Then for some $i < l$, J_i, J_l is an elliptic pair and $\langle J_i, J_l \rangle$ is nilpotent by Lemma 3.2. Now $\langle J_i, J_l \rangle = \langle x_1, \cdots, x_{r-2}, x_{r-1}, x_r^i, x_r^k \rangle$ where $k = l - i \neq 0$. Same argument as in the case $r = 2$ now shows that G is hypercentral and hence nilpotent. This completes the proof.

References

[1]   M. Curzio, J. Lennox, A. Rhemtulla and J. Wiegold, 'Groups with many permutable subgroups', to appear.

[2]   J.H. Evertse, 'On sums of S-units and linear recurrences', *Compositio Math.* 53(1984), 225-244.

[3]   J.R.J. Groves, 'A conjecture of Lennox and Wiegold concerning supersoluble groups', *J. Austral. Math. Soc.* 35(1983), 218-220.

[4]   P.H. Kropholler, 'On finitely generated soluble groups with no large wreath product sections', *Proc. London Math. Soc.* (3) 49(1984), 155-169.

[5]   J. Lennox and J. Wiegold, 'Extensions of a problem of Paul Erdös on groups', J. Austral. Math. Soc. 31(1981), 459-463.

[6]   A. Lubotzky, 'A group theoretic characterization of linear groups', *J. Algebra* 113(1988), 207-214.

[7]   A. Lubotzky and A. Mann, 'Powerful p-groups I and II', *J. Algebra* 105(1987), 484-515.

[8]   B.H. Neumann, 'A problem of Paul Erdös on groups', *J. Austral. Math. Soc.* 21(1976), 467-472.

[9]   A.H.Rhemtulla and J.S. Wilson, 'On elliptically embedded subgroups of soluble groups', *Canad. J. Math.* 39(1987),956-968.

[10] J.Tits, 'Free subgroups in linear groups', *J. Algebras* 20 (1972), 250-270.

[11] A.J. Van der Poorten, 'Additive relations in number fields', *Séminaire de Théorie des Nombres de Paris* (1982-1983), 259-266.

University of Alberta
Edmonton, T6G 2G1
Canada

REFLECTIONS ON THE CONSTRUCTIVE
THEORY OF POLYCYCLIC GROUPS

Derek J. S. Robinson

Introduction.

A group is said to be *polycyclic* if it has a series of finite length with cyclic factors. Polycyclic groups were introduced fifty years ago by K. A. Hirsch in a well-known paper [4] which marked the beginning of the theory of infinite soluble groups. Since that time polycyclic groups have been studied by many well-known names in group theory, among them A. I. Mal'cev and P. Hall.

Numerous characterizations of polycyclic groups are known. We mention two of them: (i) soluble groups satisfying max, the maximal condition on subgroups, and (ii) soluble subgroups of the groups $GL(n, \mathbf{Z})$. An excellent reference for polycyclic groups is the book of D. Segal [15].

Recall a basic fact about polycyclic groups due to Hall: every polycyclic group is finitely presented (see [12] or [15]). This allows the possibility of what we shall call the constructive theory of polycyclic groups. The object of this theory is to exhibit certain partial recursive functions which, when a polycyclic group is given by means of a finite presentation, provide information about the group. We shall not give a formal definition of a partial recursive function, but will content ourselves with stating that these are the functions whose values can be computed by a Turing machine (for more on this see [1], [14]). Thus, roughly speaking, the constructive theory is concerned with information about a polycyclic group which could, in principle at least, be obtained by programming a computer.

It is remarkable that little has been published on the constructive theory of polycyclic groups. Apart from the solutions of the standard decision problems and the isomorphism problem, we are aware of only the article of Baumslag, Cannonito and Miller [2] and a fragmentary, unpublished dissertation of N. Maxwell.

Our aim is to survey some of the things that can be done constructively for polycyclic groups. Proofs will generally be omitted unless very short. Details will appear in a forthcoming joint work with G. Baumslag, F. B. Cannonito and D. Segal.

Results.

We begin by recalling that the three classical decision problems are soluble for polycyclic groups.

In what follows G always denotes a polycyclic group with a given finite presentation in generators $x_1, x_2, \ldots, x_n$. It is understood that all algorithms are uniform in the presentation. Most of the algorithms are valid for a polycyclic-by-finite groups.

(1) (The word problem). *There is an algorithm to decide if a word w in x is the identity in G.*

We shall describe the algorithm; its effectiveness rests on the fact that polycyclic groups are residually finite, a result of Hirsch [5]. There are two procedures involved; the first calls for the enumeration of all finite groups, say by means of their group tables, the construction of the finitely many homomorphisms of G in each finite group, and a check to see if the image of w is trivial in each finite image of G. If this is not so, the procedure stops. The second procedure simply requires the enumeration of all relators, by listing all consequences of the given finite set of relators, and then searches for the word w among these relators. If w appears, the procedure stops. By residual finiteness either the first procedure stops and $w \neq 1$ or the second stops and $w = 1$ in G.

Two comments on this algorithm are in order. Firstly it clearly applies to any finitely presented residually finite group. Secondly this is typical of the sort of algorithm that would be rejected by a strict constructionist on the grounds that we cannot a *a priori* give a bound for the number of steps before an answer is obtained.

(2) (The generalized word problem). *There is an algorithm which, when words $w, w_1, \ldots, w_m$ are given, decides if $w \in \langle w_1, \ldots, w_m \rangle$ in G.*

One proof of this hinges on the theorem of Mal'cev that every subgroup of a polycyclic group is closed in the profinite topology (for an elementary proof due to J. S. Wilson see [12]). Thus one can simply check if $w \in \langle w_1, \ldots, w_m \rangle$ in finite quotients of G, and at the same time list the elements of the subgroup $\langle w_1, \ldots, w_m \rangle$ of G and look for w, (using solubility of the word problem in G). For a different approach see [2].

(3) (The conjugacy problem). *There is an algorithm to decide if two words w_1, w_2 are conjugate as elements of G.*

This depends on the fact that polycyclic groups are *conjugacy separable*, i.e., if two elements of a polycyclic group are not conjugate, they must fail to be conjugate in some finite quotient. (Formanek [3], Remeslennikov [10]). This proof of this theorem involves some non-trivial algebraic number theory and it cannot be considered an elementary result: for a clear account see [15].

We mention in passing the *isomorphism problem* for polycyclic groups: is there an algorithm to decide if two given polycyclic groups are isomorphic? For the current position on this see [15] and [16].

Constructing subgroups.

For brevity let us say that we can *find* a subgroup H of G if there is a uniform algorithm which produces a finite set of generators for H. The word "find" will be used in the same sense in slightly different situations.

We begin with a simple but basic result.

(4) *Given finite subsets* $U = \{u_1, \ldots, u_\ell\}$ *and* $V = \{v_1, \ldots, v_m\}$, *we can find the normal closure* $\overline{U}$ *of* U *in* $\langle U, V \rangle$.

Proof. Define an ascending chain of finite subsets U_i by the rules $U_0 = U$ and

$$U_{i+1} = U_i \cup \bigcup_{j=1}^{m} \left(U_i^{v_j} \cup U_i^{v_j^{-1}} \right).$$

Write $H_i = \langle U_i \rangle$, so that $H_1 \leq H_2 \leq \cdots$ and $\overline{U} = \bigcup_i H_i$. Now because of the maximal condition $H_i = H_{i+1}$ for some i, and thus $H_i = \overline{U}$. Moreover we can detect this i because $H_i = H_{i+1}$ if and only if $U_i^{v_j} \cup U_i^{v_j^{-1}} \subseteq H_i$ for $j = 1, 2, \ldots, m$; by (2) we can decide if this holds.

Corollary. *We can find the terms of the derived series and the lower central series of* G. *Hence we can find a series with cyclic factors in* G.

For example, if $U = \{[x_i, x_j] \mid i < j = 1, \ldots, n\}$, then we can find $\langle U^G \rangle = G'$.

(5) *Given a finite subset* U *of* G *we can decide if* $H = \langle U \rangle$ *is subnormal in* G.

Proof. Recall the following theorem of Kegel [7]; a subgroup H is subnormal in G if and only if HN/N is subnormal in G/N for all finite quotients G/N. Thus our first procedure searches through the finite quotients of G and checks to see if the image of H is subnormal, stopping if this ever fails to be the case. The second procedure uses (4) to find the successive normal closures H_i of H in G, i.e. $H_1 = \langle H^G \rangle$, $H_2 = \langle H^{H_1} \rangle$, etc., and then for each successive i decides if $H_i = H$ by using (2). The procedure stops if H_i equals H.

(6) **(Fundamental theorem on subgroups).** *Let* G *be a polycyclic group given by a finite presentation. Then there is a uniform algorithm which, when given a finite subset* U *of* G, *produces a finite presentation of* $\langle U \rangle$.

This is of course the key to the structure of the subgroups of G. It may well be known; certainly for finitely generated nilpotent groups it appears in [6]. The proof involves the use of polycyclic presentations which are defined below.

The theorem has many applications, for example

(7) *If an endomorphism θ of G is given by its effects on the generators, we can find $\operatorname{Im}\theta$ and $\operatorname{Ker}\theta$.*

Of course a finite set of generators of $\operatorname{Im}\theta$ is already at hand. By (6) we can find a finite presentation of $\operatorname{Im}\theta \simeq G/\operatorname{Ker}\theta$, and hence a finite set of generators for $\operatorname{Ker}\theta$ as a normal subgroup of G. Now apply (4).

Polycyclic presentations.

A *polycyclic presentation* of a group G is a finite presentation with generators x_1, x_2, ..., x_n and relations

$$\left.\begin{array}{l} x_i^{x_j} = v_{ij}(x_1,\ldots,x_{j-1}), \\ x_i^{x_j^{-1}} = v'_{ij}(x_1,\ldots,x_{j-1}), \end{array}\right\} \quad 1 \leq i < j \leq n,$$

and

$$x_i^{e_i} = u_i(x_1,\ldots,x_{i-1}), \quad 1 \leq i \leq n,$$

where v_{ij}, v'_{ij}, u_i are certain words and $1 < e_i \leq \infty$, the third relation being vacuous if $e_i = \infty$.

Let $G_i = \langle x_1,\ldots,x_i\rangle$; then there is a series $1 = G_0 \triangleleft \cdots \triangleleft G_n = G$ and G_i/G_{i-1} is cyclic of order dividing e_i. Call the presentation *consistent* if it induces a presentation of each G_i in x_1, ..., x_i (in which case $|G_i : G_{i-1}| = e_i$). Clearly a group has a (consistent) polycyclic presentation if and only if it is a polycyclic group. (The above is a minor variation of the definition in Baumslag, Cannonito and Miller [2]; these authors use the term "honest" instead of "consistent").

(8) *If a polycyclic group G is given by a finite presentation, there is a uniform procedure to produce a consistent polycyclic presentation.*

Proof. Apply sequences of Tietze transformations to the given presentation until a polycyclic presentation appears, and then test for consistency using [2], Theorem 4.3.

Corollary. *We can compute the Hirsch number $h(G)$ of G.*

For this is the number of infinite e_i's in a consistent polycyclic presentation.

One would expect consistent polycyclic presentations to be the most convenient setting for practical algorithms for polycyclic groups; we understand that Charles Sims has already produced some algorithms of this type.

A basic role is played in the construction of subgroups by the following result.

(9) *Given finite presentations of a polycyclic group G and an abelian group A, together with an explicit G-module structure for A (defined in terms of the generators of the two presentations), we can find $C_G(A)$.*

The proof depends on an algorithm from commutative algebra, for example, an algorithm to find a finite presentation of the unit group of an algebraic number field (see [9] and [17]).

Using (9) one can easily prove the following results.

(10) *We can find the centre of a polycyclic group.*

(11) *We can find the Fitting subgroup of a polycyclic group.*

The basis for this last result (which also appears in the unfinished dissertation of N. Maxwell) is the following well-known fact. A polycyclic group G has a series of normal subgroups $1 = G_0 < G_1 < \cdots < G_m = G$ such that G_{i+1}/G_i is either a chief factor or else is a torsion-free, G-rationally irreducible abelian group; then the Fitting subgroup of G is the intersection of all the $C_G(G_{i+1}/G_i)$.

(12) *Given finite subsets X, Y of a polycyclic group G, we can find $\langle X \rangle \cap \langle Y \rangle$.*

This is more difficult to establish since neither subgroup need be normal. A double induction on $h(G)$ and $h(G) - h(H)$ leads to a reduction to a semidirect product, which can be handled directly.

Corollary. *We can find the core of $\langle X \rangle$ in G.*

This follows from (12) and the theorem of Rhemtulla [11] that the core of a subgroup of a polycyclic group is the intersection of finitely many of its conjugates.

(13) *Given finite subsets X, Y of a polycyclic group G we can find $C_{\langle Y \rangle}(X)$.*

This too involves a reduction to a semidirect product.

(14) *We can find the FC-centre of a polycyclic group G.*

(Recall that the FC-centre is the subgroup of all elements with finitely many conjugates).

The most difficult subgroup to construct to date has proved to be the Frattini subgroup.

(15) *We can find the Frattini subgroup $\varphi(G)$ of a polycyclic group G.*

The idea here is to reduce to the case where $G = Q \ltimes A$, with Q, A both free abelian. Then $\varphi(G) = \varphi_Q(A)$, the intersection of the maximal ZQ-submodules of A. The rest is constructive commutative algebra.

A polycyclic group will generally have infinitely many elements of finite order, so there can be no question of finding them all. However, according to a theorem of Mal'cev [8], the finite subgroups fall into finitely many conjugacy classes. This is the basis for the next result.

(16) *Given a polycyclic group G one can find a finite set of finite subgroups such that every finite subgroup of G is conjugate to one of these.*

Corollary. *We can decide if G is torsion-free.*

Finally a result which is in the spirit of [13].

(17) *Given an automorphism of a polycyclic group G by its effect on the generators, there is a uniform procedure to decide if it is an inner automorphism.*

There are certainly many other natural questions that can be raised in the constructive theory of polycyclic groups. One that may prove difficult is to find the minimum number of generators of a polycyclic group.

References

[1] M.A. Arbib, *Theories of Abstract Automata*, Prentice Hall, Englewood Cliffs, NJ (1969).

[2] G. Baumslag, F.B. Cannonito and C.F. Miller III, 'Infinitely generated subgroups of finitely presented groups I.', *Math. Z.* **153** (1979), 117–134.

[3] E. Formanek, 'Conjugacy separability in polycyclic groups', *J. Algebra* **42** (1976), 1–10.

[4] K.A. Hirsch, 'On infinite soluble groups I', *Proc. London Math. Soc.* (2) **44** (1938), 53–60.

[5] K.A. Hirsch, 'On infinite soluble groups IV', *J. London Math. Soc.* **27** (1952), 81–85.

[6] M.I. Kargapolov, V.N. Remeslennikov, V.A. Roman'kov, V.A. Čurkin, 'Algorithmic problems for σ-power groups', *Algebra i Logika* **8** (1969), 643–659.

[7] O.H. Kegel, 'Über den Normalisator von subnormalen und erreichbaren Untergruppen', *Math. Ann.* **163** (1966), 248–258.

[8] A.I. Mal'cev, 'On certain classes of infinite soluble groups, *Mat. Sb.* **28** (1951), 567–588 = A.M.S. Transl. (2) **2** (1956), 2–21.

[9] M. Pohst and H. Zassenhaus, 'An effective number geometric method of computing the fundamental units of an algebraic number field', *Math. Comp.* **31** (1977), 754–774.

[10] V.N. Remeslennikov, 'Conjugacy in polycyclic groups', *Algebra i Logika* **8** (1969), 712–725.

[11] A.H. Rhemtulla, 'A minimality property of polycyclic groups', *J. London Math. Soc.* **42** (1967), 456–462.

[12] D.J.S. Robinson, *A Course in the Theory of Groups*, Springer, Berlin, (1982).

[13] D.J.S. Robinson, 'Algorithmic problems for automorphisms and endomorphisms of infinite soluble groups', in *Algebraic Structures and Number Theory*, Lecture Notes in Math., to appear.

[14] H. Rogers, *Theory of Recursive Functions and Effective Computability*, McGraw-Hill, New York, (1967).

[15] D. Segal, *Polycyclic Groups*, Cambridge, (1983).

[16] D. Segal, 'The general polycyclic group', *Bull. London Math. Soc.* **19** (1987), 49–56.

[17] H. Zassenhaus, 'On the units of orders', *J. Algebra* **20** (1972), 368–375.

Department of Mathematics
University of Illinois at Urbana-Champaign
1409 West Green Street
Urbana, Illinois 61801
U. S. A.

MINIMAL GENERATING SYSTEMS FOR PLANE DISCONTINUOUS GROUPS AND AN EQUATION IN FREE GROUPS

Gerhard Rosenberger

0. Introduction.

Let be given a plane discontinuous group

$$G = \langle s_1, \cdots, s_m, a_1, \cdots, a_p \mid s_1^{\gamma_1} = \cdots = s_m^{\gamma_m} = s_1 s_2 \cdots s_m P(a_1, \cdots, a_p) = 1 \rangle$$

with all $\gamma_i \geq 2$, which satisfies either

(a) $P(a_1, \cdots, a_p) = [a_1, a_2] \cdots [a_{p-1}, a_p]$, p even, and

    (i) $p \geq 2$ or

    (ii) $m \geq 4$ or

    (iii) $p = 0$, $m = 3$ and $\frac{1}{\gamma_1} + \frac{1}{\gamma_2} + \frac{1}{\gamma_3} \leq 1$, or

(b) $P(a_1, \cdots, a_p) = a_1^2 \cdots a_p^2$ and

    (i) $p \geq 2$ or

    (ii) $p = 1$ and $m \geq 2$.

In [10] I proved the following result. If $p \geq 2$, $m \geq 2$ and $\{x_1, \cdots, x_{p+m-1}\}$ is a generating system of G then $\{x_1, \cdots, x_{p+m-1}\}$ is Nielsen equivalent to a system $\{s_{\nu_1}^{\beta_1}, \cdots s_{\nu_{m-1}}^{\beta_{m-1}}, a_1, \cdots, a_p\}$ with $\nu_i \in \{1, \cdots, m\}$, $\nu_1 < \cdots < \nu_{m-1}$ (if $m \geq 3$) and $1 \leq \beta_i < \gamma_{\nu_i}$, $(\beta_i, \gamma_{\nu_i}) = 1$.

A slightly different, more compact proof is published in [16]. This proof makes use of a result about an equation in a free group [13] which is based on the interpretation of a theorem of G. Baumslag and Steinberg [2] by B. Baumslag and Levin [1]. I want to mention that the proof in [10] does not depend on the result in [13] (see also [11]). Unfortunately the interpretation by B. Baumslag and Levin is not correct. In this paper we first give a counterexample to this interpretation which is also a counterexample to Theorem 1 of [13] because the reduction argument from a proper power to the root of a word in a free group does not work. Afterwards we give a correction of Theorem 1 of [13] which is good enough to give a direct correction of the proof in [16]. Finally we extend the result of [16]. Such an extension is of interest because the Heegard splittings of a Seifert fibered 3-manifold with minimal genus are closely related to the Nielsen equivalence classes of minimal generating systems for the underlying Fuchsian group (see [3] and [5]).

1. Preliminary remarks.

In this paper we use the terminology and notation of [7] and [22]; here $\langle \cdots \mid \cdots \rangle$ indicates a description of a group in terms of generators and relations. In a group G an element x is called a *proper power* if there is an $y \in G$ such that $x = y^\gamma$ for some $|\gamma| \geq 2$. By $\langle a_1, \cdots, a_n \rangle$ we denote the group generated by $a_1, \cdots, a_n$; $[a, b] = aba^{-1}b^{-1}$ means the commutator of $a, b \in G$ (G a group), (n, m) the greatest common divisor and $n|m$ that n divides m ($n, m \in \mathbf{N}$).

Frequently we obtain from one system $\{x_1, \cdots, x_m\}$ a new one by Nielsen transformations and then denote the latter by the same symbols. Let $G = \overset{k}{\underset{i=1}{*_A}} H_i$ be the free product of $H_1, \cdots, H_k$ with amalgam $A = H_i \cap H_j$, $i \neq j$; if $A = \{1\}$ then we have the case of the free product of $H_1, \cdots, H_k$. We assume that in G a length L and an order are introduced as in [4], [17] and [20]. For a finite system we always may choose a suitable order such that this system is Nielsen equivalent to a Nielsen reduced system. From [4], [17] and [20] we obtain the following lemma:

Lemma (1.1): *Let $\{x_1, \cdots, x_m\} \subset G$ be a finite, Nielsen reduced system of elements of G. Then one of the following cases holds:*

(1.2) *For every $w \in \langle x_1, \cdots, x_m \rangle$ there is a presentation $w = \prod_{i=1}^{q} x_{\nu_i}^{\varepsilon_i}$, $\varepsilon_i = \pm 1$, $\varepsilon_i = \varepsilon_{i+1}$ if $\nu_i = \nu_{i+1}$ with $L(x_{\nu_i}) \leq L(w)$ for $i = 1, \cdots, q$.*

(1.3) *There is a product $a = \prod_{i=1}^{q} x_{\nu_i}^{\varepsilon_i}$, $a \neq 1$, with $x_{\nu_i} \in A$ ($i = 1, \cdots, q$), and in one of the factors H_j there is an element $x \notin A$ with $xax^{-1} \in A$.*

(1.4) *Of the x_i there are p, $p \geq 1$, contained in a subgroup of G conjugate to some H_j, and   a certain product of them is conjugate to a non-trivial element of A.*

(1.5) *There is a $g \in G$ such that some $x_i \notin gAg^{-1}$, but for a suitable $k \in \mathbf{N}$ we have*
$$x_i^k \in gAg^{-1}. \qquad \qquad \square$$

Remark: If $\{x_1, \cdots, x_m\}$ is a generating system of G, then in case (1.4) we find that $p \geq 2$, because then conjugations determine a Nielsen transformation. If we are interested mainly in combinatorial description of $\langle x_1, \cdots, x_m \rangle$ in terms of generators and relations, we find again that $p \geq 2$ in case (1.4), possibly after a suitable conjugation. This remark implies a certain correction of Theorem 1 of [20]. The correct form of Theorem 1 of [20], as stated, is obtained by replacing the strict inequality $<$ in (2.4 (d)) by a weak inequality $\leq$.

2. An equation in a free group.

In [2] G. Baumslag proved

Theorem (2.1) [2]: *Let $w = w(x_1, \cdots, x_n)$ be an element of a free group F freely generated by $x_1, \cdots, x_n$ which is neither a proper power nor a primitive. If $g_1, \cdots, g_n$, g are elements of a free* *group connected by the relation*

$$w(g_1, \cdots, g_n) = g^m \quad (m > 1)$$

then the rank of the group generated by $g_1, \cdots, g_n, g$ *is at most* $n - 1$.   □

Theorem (2.1) was independently proved by Steinberg (see [2]). B. Baumslag and Levin interpreted Theorem (2.1) to the following statement (Lemma 2 in [1]): Let $w(x_1, \cdots, x_n)$ be a reduced word which is not primitive in the free group freely generated by $x_1, \cdots, x_n$. Suppose that in a free group G, $g_1, \cdots, g_n$ freely generate a group of rank n and that g in G is such that $g^m = w(g_1, \cdots, g_n)$ where m is an integer greater than 1. Then $w(x_1, \cdots, x_n)$ is a proper power.   □

They stated this (without proof) as a direct consequence of Theorem (2.1). But the interpretation of Theorem (2.1) by B. Baumslag and Levin is not correct as the following example shows.

Let $F = \langle a, b \; ; \; \rangle$ be the free group on a and b. Let $x = aba^{-1}$, $y = bab^{-1}$, $u = a^2$, $v = b^2$ and H be the subgroup generated by u, v, x and y. The normal subgroup of F generated by a^2, b^2 and $(ab)^3$ is contained in H (we remark that $xuyv = (ab)^3$). This shows that H has index 3 in F. If we regard F as a Fuchsian group with finite volume then the Riemann-Hurwitz Relation gives that H is a free group of rank four, that means, $\{x, y, u, v\}$ is a free generating system of H.

Hence, H is also free on $x_1 = x$, $x_2 = v^{-1}yvu^{-1}$, $x_3 = v^{-1}yv$ and $x_4 = v$; $[x_1, x_2][x_3, x_4]$ is reduced (with respect to the free generating system $\{x_1, x_2, x_3, x_4\}$) and not primitive in H. In H we have the relation $[x_1, x_2][x_3, x_4] = [a, b]^3$, but $[x_1, x_2][x_3, x_4]$ is not a proper power in H.

This example, clearly, is not a counterexample to Theorem (2.1), only to the interpretation in Lemma 2 of [1].

Unfortunately, I used this interpretation in the reduction argument from $\alpha > 1$ to $\alpha = 1$ in the proof of Theorem 1 in [13]. In the first version of [13] I proved a much weaker result which was good enough for the applications in [13]. In a letter the editor wrote me that the referee has the opinion that I should try to prove a more general result. At the same time I saw Lemma 2 in [1] together with the remark that it was proved in [2]. Unfortunately I did not look into the paper [2] at that time and made straightforward use of its interpretation in Lemma 2 of [1]. Because of a delay in the publication of [12] I also used this interpretation to give an extension of the earlier versions of Theorem 1 and Theorem 2 of [12]. The above example, clearly, is also a counterexample to Theorem 1 of [13]. The proof in [13] now only gives the result for $\alpha = 1$, that means the following.

Theorem (2.2) [13]: *Let* F *be the free group on* $a_1, \cdots, a_p$ $(p \geq 1)$ *and*

$$P(a_1, \cdots, a_p) := a_1^{\alpha_1} \cdots a_n^{\alpha_n} [a_{n+1}, a_{n+2}] \cdots [a_{p-1}, a_p] \in F$$

an alternating product in F *with* $0 \leq n \leq p$, $p - n$ *even and* $\alpha_i \geq 1$ *for* $i = 1, \cdots, n$. *Let* $\{x_1, \cdots, x_m\}$ $(m \geq 1)$ *be any system in* F *and* X *be the subgroup of* F *generated by* $x_1, \cdots, x_m$. *Let* $y^{-1} P(a_1, \cdots, a_p) y \in X$ *for some* $y \in F$. *Then one of the following cases occurs.*

(a) $\{x_1, \cdots, x_m\}$ is Nielsen equivalent to a system $\{y_1, \cdots, y_m\}$ with $y_1 = zP(a_1, \cdots, a_p)z^{-1}$, $z \in F$.

(b) We have $m \geq p$, and $\{x_1, \cdots, x_m\}$ is Nielsen equivalent to a system $\{y_1, \cdots, y_m\}$ with $y_i = za_i^{\gamma_i}z^{-1}$, $1 \leq \gamma_i < \alpha_i$, $\gamma_i | \alpha_i$ $(i = 1, \cdots, n)$, $y_j = za_jz^{-1}$ $(j = n+1, \cdots, p)$ and $z \in F$. $\qquad\square$

For the weaker statements in Theorem 1 and Theorem 2 of [12] we have the corresponding alternation. These alternations are good enough for the proof of Theorem 3 and 4 in [12] and for the proof of Theorem 2 in [13], especially, for the proof of the conditions for a system $\{x_1, \cdots, x_n\}$ to be a free generating system in a free group of rank n. But these alternations are not good enough for the proof of the other applications in [13], that means, for the proof of Theorem 3 to 5 in [13] about generating systems (see Lemma (1.1) and the introduction). For this we need a result for proper powers of $P(a_1, \cdots, a_p)$, but, clearly, we only need the assumption $1 \leq m \leq p$. Before giving such a result we prove the following.

Lemma (2.3): *Let F be the free group on $a_1, \cdots, a_p$ $(p \geq 1, p$ even$)$ and*

$$P(a_1, \cdots, a_p) = [a_1, a_2] \cdots [a_{p-1}, a_p] \in F.$$

Let $\{x_1, \cdots, x_m\}$ $(1 \leq m \leq p)$ be any system in F. Then there does not hold any equation

$$(2.4) \qquad \prod_{i=1}^{q} x_{\nu_i}^{\varepsilon_i} = P^\alpha(a_1, \cdots, a_p), \quad \varepsilon_i = \pm 1,$$

$\varepsilon_i = \varepsilon_{i+1}$ *if* $\nu_i = \nu_{i+1}$, $\alpha \geq 2$, *such that each x_j is conjugate to a proper power of some a_k or occurs in (2.4) exactly once with exponent $+1$ and once with exponent -1.*

Proof. Suppose such an equation (2.4) holds. We first handle the case $p = 2$. Let $p = 2$, then obviously $m = 2$. If (after a suitable renumeration) x_1 is conjugate to a_1^δ, $|\delta| \geq 2$, then we introduce the relation $a_1^\delta = 1$ and get $a_i^\gamma = z[a_1, a_2]^\alpha z^{-1}$, $i = 1$ or 2, for some $\gamma \in \mathbb{Z}$ and $z \in \langle a_1, a_2 \mid a_1^\delta = 1 \rangle$ which gives a contradiction.

If no x_j is conjugate to a proper power of some a_k then $[a_1, a_2]^\alpha$ is a commutator in F which gives a contradiction by Theorem (2.1). Hence, if $p = 2$ then such an equation (2.4) does not hold.

From now on let $p = 2n$, $n \geq 2$, and Lemma (2.3) be proved for all $p_1 = 2n_1$ with $n_1 < n$.

We first show that there is no primitive element in $X = \langle x_1, \cdots, x_m \rangle$ which is conjugate to a proper power of $P(a_1, \cdots, a_p)$. Suppose there is such a primitive element. If two of the x_i are conjugate to a proper power of the same a_j, say a_p, we introduce the relation $a_p = 1$ which gives $m - 2 \leq p - 2$ and an equation like (2.4) in the free group on $a_1, \cdots, a_{p-2}$ which contradicts our induction hypothesis. Hence there are no two x_i which are conjugate to a proper power of the same a_j. From this we get that $P^\alpha(a_1, \cdots, a_n)$ must be contained in the commutator subgroup X' of X which gives a contradiction because in a free group no nontrivial power of a primitive element is contained in its commutator subgroup.

Therefore, no primitive element in X is conjugate to a power of $P(a_1, \cdots, a_p)$. In F we introduce the free length L and a suitable lexicographical order with respect to the generating system $\{a_1, \cdots, a_p\}$. With help of the classical Nielsen reduction method we obtain from $\{x_1, \cdots, x_m\}$ a system $\{y_1, \cdots, y_k\}$, $1 \leq k \leq m \leq p$, which generates X freely and has the Nielsen property, that means, no y_i and no y_i^{-1} cancels more than half of another y_j, no two elements cancel another one completely. If we write each y_i^ε, $\varepsilon = \pm 1$, as a freely reduced word in a_j, a_j^{-1} then for each y_i^μ, $\mu = \pm 1$, there is a symbol which remains unchanged in each freely reduced word in $y_1, \cdots, y_k$ at the place where y_i^μ occurs. For each y_i we choose such a symbol and take the inverse symbol for y_i^{-1}. We have an equation

$$(2.5) \quad \prod_{i=1}^{t} y_{\nu_i}^{\varepsilon_i} = P^\alpha(a_1, \cdots, a_p), \quad \varepsilon_i = \pm 1, \ \varepsilon_i = \varepsilon_{i+1} \text{ if } \nu_i = \nu_{i+1}.$$

If a factor $y_{\nu_i}^{\varepsilon_i}$ occurs twice in (2.5) then we have a partial product $\prod_{i=e}^{s} y_{\nu_i}^{\varepsilon_i}$ with $\nu_e = \nu_{s+1}$, $\varepsilon_e = \varepsilon_{s+1}$ and no $y_{\nu_i}^{\varepsilon_i}$ occurs twice in $y_{\nu_e}^{\varepsilon_e}, \cdots, y_{\nu_s}^{\varepsilon_s}$.

The symbol of $y_{\nu_e}^{\varepsilon_e}$ occurs twice in $\prod_{i=1}^{t} y_{\nu_i}^{\varepsilon_i}$ and, therefore, also twice in $P^\alpha(a_1, \cdots, a_p)$. Hence, $\prod_{i=e}^{s} y_{\nu_i}^{\varepsilon_i}$ is conjugate to a power of $P(a_1, \cdots, a_p)$. Therefore, after a suitable conjugation, we obtain an equation

$$(2.6) \quad \prod_{i=1}^{r} y_{\nu_i}^{\varepsilon_i} = P^\beta(a_1, \cdots, a_p), \quad \beta \geq 1, \ \varepsilon_i = \pm 1, \ \varepsilon_i = \varepsilon_{i+1} \text{ if } \nu_i = \nu_{i+1},$$

in which each y_{ν_i} occurs exactly once with exponent $+1$ and once with exponent -1 (if we abelianize then we see that each y_{ν_i} occurs as much as $y_{\nu_i}^{-1}$. We want to show that $\beta = 1$. Suppose $\beta \geq 2$.

Let h be the number of the y_i which occur in (2.6). Without any loss of generality let $y_1, \cdots, y_h$ occur in (2.6). We regard the subgroup $H = \langle y_1, \cdots, y_h, \ P(a_1, \cdots, a_p) \rangle$ of F. By theorem (2.1) the rank of H is at most $h - 1$ Also no conjugate of $P(a_1, \cdots, a_p)$ is a primitive element in H because $P^\beta(a_1, \cdots, a_p)$ is contained in the commutator subgroup H'. This means $H = F$ by theorem (2.2) and, hence, $p \leq h - 1 \leq p - 1$ because $h \leq k \leq m \leq p$. This gives a contradiction. Hence, $\beta = 1$. Now, by Theorem (2.2), $m = p$ and $\{x_1, \cdots, x_m\}$ is a free generating system of F and is Nielsen equivalent to $\{a_1, \cdots, a_p\}$ which gives a contradiction (an equation (2.4) obviously cannot hold if $\{x_1, \cdots, x_m\}$ is a free generating system of F).

Therefore, altogether, such an equation (2.4) does not hold in F. $\qquad\qquad \square$

The Nielsen method in free groups, when taken together with the combination of Theorem (2.1) and Theorem (2.2) as in the proof of Lemma (2.3), now straightforward gives the following (see also [18]).

Theorem (2.7): *Let F be a free group on $a_1, \cdots, a_p$ ($p \geq 1$, p even) and $P(a_1, \cdots, a_p) = [a_1, a_2] \cdots [a_{p-1}, a_p] \in F$. Let $\{x_1, \cdots, x_m\}$ ($1 \leq m$) be any system in F and X be the subgroup of F generated by $x_1, \cdots, x_m$. Let $y^{-1} P^\alpha(a_1, \cdots, a_p) y \in X$ for some $\alpha \neq 0$ and $y \in F$. Then one of the following cases occurs:*

(a) $\{x_1, \cdots, x_m\}$ is Nielsen equivalent to a system $\{y_1, \cdots, y_m\}$ with $y_1 = z P^\beta(a_1, \cdots, a_p) z^{-1}$,

$\beta \geq 1$ and $z \in F$, and β is the smallest positive number for which a relation
$\quad y^{-1}P^{\beta}(a_1, \cdots, a_p)y \in X$ holds for some $y \in F$.

(b) X has finite index $\beta = [F : X]$ in F, where β, again, is the smallest positive number for which a relation $y^{-1}P^{\beta}(a_1, \cdots, a_p)y \in X$ holds for some $y \in F$.   □

Corollary (2.8): *Let F be the free group on a, b and $\alpha \in \mathbf{N}$, $\alpha \geq 2$. Let $x_1, y_1, \cdots, x_q, y_q \in F$ $(q \geq 1)$ such that $[a,b]^{\alpha} = \prod_{i=1}^{q}[x_i, y_i]$. Let $[a,b]^{\alpha}$ be not a proper power in the subgroup X generated by $x_1, y_1, \cdots, x_q, y_q$. Then α is odd and $q \geq \frac{\alpha+1}{2}$.*

Proof. X has finite index α in F by Theorem (2.7). If we regard F as a Fuchsian group such that $[a, b]$ is parabolic then the Riemann-Hurwitz-Relation yields to α odd and $\alpha \leq 2q - 1$ which gives the result.   □

Remarks:

1) We know from the above example that $[a, b]^3$ is a product of two commutators in the free group F on a and b.

2) Again, let F be the free group on a, b and $w \in F$, $w \neq 1$. Let $x_1, y_1, x_2, y_2 \in F$ such that $[x_1, y_1][x_2, y_2] = w^2$. Some years ago, C.C. Edmunds asked me if necessarily $[x_1, y_1] = [x_2, y_2]$. The following example gives a negative answer: $[[a, b], a][a^2, b] = [a, b]^2$. We mention that $[a, b]^2$ is a proper power in the subgroup generated by $[a, b]$, a, a^2 and b. Recently, C.C. Edmunds wrote me that J. Comerford and Y. Lee have found a similar example. We make the following conjecture: Let F and $w \neq 1$ as above. Let $x_1, y_1, x_2, y_2 \in F$ such that $[x_1, y_1][x_2, y_2] = w^{\alpha}$ for some $\alpha \geq 2$. Then $\alpha = 2$ or $\alpha = 3$.

We now present the desired result about proper powers of $P(a_1, \cdots, a_p)$.

Theorem (2.9): *Let F be the free group on $a_1, \cdots, a_p$ $(p \geq 1)$ and*

$$P(a_1, \cdots, a_p) = a_1^{\alpha_1} \cdots a_n^{\alpha_n}[a_{n+1}, a_{n+2}] \cdots [a_{p-1}, a_p] \in F$$

an alternating product in F with $0 \leq n \leq p$, $n - p$ even and $\alpha_i \geq 1$ for $i = 1, \cdots, n$. Let $\{x_1, \cdots, x_m\}$ $(1 \leq m \leq p)$ be any system in F and X be the subgroup of F generated by $x_1, \cdots, x_m$. Let $y^{-1}P^{\alpha}(a_1, \cdots, a_p)y \in X$ for some $\alpha \neq 0$ and $y \in F$. Then one of the following cases occurs:

(a) $\{x_1, \cdots, x_m\}$ is Nielsen equivalent to a system $\{y_1, \cdots, y_m\}$ with $y_1 = zP^{\beta}(a_1, \cdots, a_p)z^{-1}$, $\beta \geq 1$ and $z \in F$, and β is the smallest positive number for which a relation $\quad y^{-1}P^{\beta}(a_1, \cdots, a_p)y \in X$ holds for some $y \in F$.

(b) We have $m = p$ and $\{x_1, \cdots, x_m\}$ is Nielsen equivalent to a system $\{y_1, \cdots, y_m\}$ with $\quad y_i = za_i^{\gamma_i}z^{-1}$, $1 \leq \gamma_i < \alpha_i$, $\gamma_i|\alpha_i$ $(i = 1, \cdots, n)$, $y_j = za_jz^{-1}$ $(j = n + 1, \cdots, p)$ and $z \in F$.

Proof. In the following let α be the smallest positive number for which a relation

$y^{-1}P^\alpha(a_1, \cdots, a_p)y \in X$ holds for some $y \in F$. If $\alpha = 1$ then the statement holds by Theorem (2.2). Hence, from now on let $\alpha \geq 2$. We may assume that $y = 1$ (replace x_i by yx_iy^{-1} if necessary). We show that case (a) occurs. Suppose that case (a) does not occur. We write F as the free product $F = \langle a_1 \rangle * \cdots * \langle a_p \rangle$ of the p infinite cyclic groups $\langle a_i \rangle$. Suppose that a length L and a suitable order are introduced as in [4], [17] and [20]. We may assume that $x_i \neq 1$ for all i and that $\{x_1, \cdots, x_m\}$ is Nielsen-reduced with respect to the length L and the order (see [17]). The cases (1.3), (1.4) and (1.5) of Lemma (1.1) cannot hold. Hence, case (1.2) holds for $\{x_1, \cdots, x_m\}$. Especially, $x_1, \cdots, x_m$ freely generate X and $L(x_i^\varepsilon x_j^\eta) \geq \max\{L(x_i), L(x_j)\}$ for ε, $\eta = \pm 1$ and $i \neq j$ or $i = j$, $\varepsilon = \eta$. For the minimal α and the system $\{x_1, \cdots, x_m\}$ we have an equation

$$(2.10) \qquad \prod_{j=1}^{q} x_{\nu_j}^{\varepsilon_j} = P^\alpha(a_1, \cdots, a_p), \quad \varepsilon_j = \pm 1, \ \varepsilon_j = \varepsilon_{j+1} \text{ if } \nu_j = \nu_{j+1}.$$

Among the equations as in (2.10) we may choose one for which q is minimal. Let this be the equation (2.10). We also may assume that each x_i occurs in (2.10).

Now, we first assume that the system $\{x_1, \cdots, x_m\}$ has the Nielsen-property, that is, $L(x_k^\varepsilon x_i x_j^\eta) > L(x_k) - L(x_i) + L(x_j)$ for all $i, j, k, \varepsilon, \eta$ with $\varepsilon, \eta = \pm 1$ and $k \neq i \neq j$ or $k = i$, $\varepsilon = 1$ or $i = j$, $\eta = 1$. We assume that in (2.10) there are two indices j, i $(1 \leq j < i \leq q)$ with $\nu_j = \nu_i = k$, $\nu_h \neq k$ for $j < h < i$ and $\varepsilon_j = \varepsilon_i$. Without any loss of generality, let $\varepsilon_j = \varepsilon_i = 1$. We want to show that $j = i - 1$ and that x_k is conjugate to a power of some a_ℓ with $1 \leq \ell \leq n$. There is a stable symbol a_λ^β of x_k which can be recovered in the normal forms of all reduced words in the x_i which contain x_k ([20]). We have $x_i = ua_\lambda^\beta v$ where the normal form of u does not end and the normal form of v does not start with a power of a_λ, that is especially, $L(x_k) = L(u) + L(v) + 1$.

Let $1 \leq \lambda \leq n$. We obtain

$$x_k \Big(\prod_{\mu=j+1}^{i-1} x_{\nu_\mu}^{\varepsilon_\mu} \Big) x_k = ua_\lambda^\beta v \Big(\prod_{\mu=j+1}^{i-1} x_{\nu_\mu}^{\varepsilon_\mu} \Big) ua_\lambda^\beta v$$

$$= ua_\lambda^{\beta'} (a_\lambda^{\alpha_\lambda} \cdots a_n^{\alpha_n} [a_{n+1}, a_{n+2}] \cdots [a_{p-1}, a_p] a_1^{\alpha_1} \cdots a_{\lambda-1}^{\alpha_{\lambda-1}})^\gamma a_\lambda^{\beta''} v$$

with $\lambda \geq 0$.

Suppose $\gamma > 0$. Then $\beta' + \alpha_\lambda = \beta$, $\beta'' = \alpha_\lambda$ or $\beta' = 0$, $\beta'' = \beta$ because, if $\beta' + \alpha_\lambda \neq \beta$ then necessarily $\beta'' = \beta$ and $\beta' + \alpha_\lambda = \alpha_\lambda$ in consideration of the Nielsen-property of $\{x_1, \cdots, x_m\}$. We regard the case $\beta' + \alpha_\lambda = \beta$, $\beta'' = \alpha_\lambda$, the case $\beta' = 0$, $\beta'' = \beta$ is analogous.

Let $\beta' + \alpha_\lambda = \beta$, $\beta'' = \alpha_\lambda$. Then we obtain

$$\Big(\prod_{\mu=j+1}^{i-1} x_{\nu_\mu}^{\varepsilon_\mu} \Big) x_k = v^{-1} a_\lambda^{-\alpha_\lambda} \cdots a_1^{-\alpha_1} P^\gamma(a_1, \cdots, a_p) a_1^{\alpha_1} \cdots a_\lambda^{\alpha_\lambda} v.$$

If $0 < \gamma < \alpha$ then this equation contradicts the minimality of α and if $\gamma = \alpha$ then it contradicts the minimality of q. Hence, $\gamma = 0$, that means, $j = i - 1$ and x_k is conjugate to a power of a_λ. Suppose now $n + 1 \leq \lambda \leq p$. The analogous argument shows that again $j = i - 1$ and that x_k is conjugate to a power of a_λ. Then there are necessarily $\ell \neq k \neq h$ and $\varepsilon, \eta = \pm 1$ such that $L(x_\ell^\varepsilon x_k x_h^\eta) \leq L(x_\ell) - L(x_k) + L(x_h)$ because of the special form of the equation (2.10). This contradicts the Nielsen-property of $\{x_1, \cdots, x_m\}$. Hence, $n + 1 \leq \lambda \leq p$ does not occur. Because case (a) does not occur by supposition we obtain therefore that each x_i is conjugate to a power of some a_j with $1 \leq j \leq n$ or it occurs in (2.10) exactly once with exponent $+1$ and once with exponent -1. For each a_j with $1 \leq j \leq n$ there is at least one x_i such that x_i is conjugate to a power of a_j because $P^\alpha(a_1, \cdots, a_p) \in X$ and the abelianized group F/F' is free abelian of rank p. If $n < p$ then this forces $\alpha = 1$ by Lemma (2.3) which gives a contradiction (we introduce the relations $a_1 = \cdots = a_n = 1$ if necessary and recall that $m - n \leq p - n$). Hence, case (a) holds if $n < p$.

Now, let $n = p$, that is, $P(a_1, \cdots, a_p) = a_1^{\alpha_1} \cdots a_p^{\alpha_p}$. Then $m = p$ and for each a_j there is exactly one x_i which is conjugate to a power of a_j. Again we obtain $\alpha = 1$ which gives a contradiction. Hence, case (a) also holds if $n = p$.

Now, we second assume that the system $\{x_1, \cdots, x_m\}$ does not have the Nielsen-property. We assume that there are $i, j, k \in \{1, \cdots, m\}$ and $\varepsilon, \eta = \pm 1$ such that $k \neq i \neq j$ or $k = i \neq j$, $\varepsilon = 1$ or $k \neq i = j$, $\eta = 1$ and $L(x_k^\varepsilon x_i x_j^\eta) \leq L(x_k) - L(x_i) + L(x_j)$. Because $\{x_1, \cdots, x_m\}$ is Nielsen-reduced we then have the following situation (see [17] and [20]):

$$x_i = v a_\lambda^\gamma v^{-1} \text{ with } L(x_i) = 2L(v) + 1,$$

$$x_k^\varepsilon = u_k a_\lambda^\beta v^{-1} \text{ with } L(x_k) = L(v) + L(u_k) + 1,$$

$$x_j^\eta = v a_\lambda^\delta u_j \text{ with } L(x_j) = L(v) + L(u_j) + 1,$$

Suppose $\gamma = \pm 1$, say $\gamma = 1$. If $k \neq i$ then $x_k^\varepsilon x_i^{-\beta} = u_k v^{-1}$ and $L(x_k^\varepsilon x_i^{-\beta}) < L(x_k)$ which gives a contradiction because $\{x_1, \cdots, x_m\}$ is Nielsen-reduced. If $i \neq j$ then $x_i^{-\delta} x_j^\eta = v u_j$ and $L(x_i^{-\delta} x_j^\eta) < L(x_j)$ which also gives a contradiction. Hence, x_i is conjugate to a_λ^γ with $|\gamma| \geq 2$. Now, we assume that there is some x_i with the property that first $L(x_k^\varepsilon x_i x_j^\eta) > L(x_k) - L(x_i) + L(x_j)$ for all k, j, ε, η with $\varepsilon, \eta = \pm 1$ and $k \neq i \neq j$ or $k = i$, $\varepsilon = 1$ or $i = j$, $\eta = 1$ and that second x_i is not conjugate to a proper power of some a_ℓ. If this x_i is conjugate to some a_ℓ^ε, $\varepsilon = \pm 1$, say $x_i = v a_\ell^\varepsilon v^{-1}$ with $L(x_i) = 2L(v) + 1$, then there is no x_j, $j \neq i$, such that $x_j^\eta = u_j a_\ell^\varepsilon v^{-1}$, $\eta = \pm 1$, with $L(x_j) = L(u_j) + L(v) + 1$ and no x_k, $k \neq i$, such that $x_k^\eta = v a_\ell^\beta u_k$, $\eta = \pm 1$, with $L(x_k) = L(u_k) + L(v) + 1$. Hence, in any case, for this x_i we may apply the above arguments correspondingly (we may regard blocks $x_i^{\delta\nu}$ if x_i is conjugate to some a_ℓ^ε, $\varepsilon = \pm 1$) to get especially, that this x_i is conjugate to a power of some a_j with $1 \leq j \leq n$ if it occurs twice in (2.10) with the same exponent.

Hence, altogether we obtain that each x_i is conjugate to a power of some a_j with $1 \leq j \leq n$ or it is conjugate to a proper power of some a_j with $n + 1 \leq j \leq p$ or it occurs in (2.10) exactly once with exponent $+1$ and once with exponent -1. Again, for each a_j with $1 \leq j \leq n$ there is at least one x_i such that x_i is conjugate to a power of a_j. We may apply the above arguments to get $\alpha = 1$ which gives a contradiction. Hence, case (a) holds. $\square$

Remarks:

1) Theorem 1 of [13] was also mentioned in the survey article [14; Theorem (2.3)], therefore, we should replace Theorem (2.3) in [14] by the above Theorems (2.2) and (2.9). These theorems, clearly, are good enough for the proof of Theorem (2.4) in [14]. They are also good enough for the proof of Theorem (2.2) in [10] if we replace Theorem (1.3) in [16] (which is identical with Theorem 2 in [13]) by the above Theorems (2.2) and (2.9); we remark that Corollaries (1.4) and (1.5) in [16] also follow from these theorems with the same proof. Finally, Theorems (2.2) and (2.9) are good enough for the proof of Theorems 3 to 5 in [13].

2) The interpretation of [2] in Lemma 2 of [1] was used in two more papers, namely in [15] and [18].
In [18] it was used for the proof of Theorem (2.12). Analogously as in [13] this proof now only gives
the result for $\alpha = 1$, that means, we have to replace $\alpha \neq 0$ by $\alpha = 1$ in Theorem (2.12) of [18]. But
with this alternation this theorem is just a special case of the main Theorem (2.1) in [18].

Theorem (2.11) [18]: *Let $G = H_1 * \cdots * H_n$ ($n \geq 2$) be the free product of the groups $H_1, \cdots, H_n$.
Let $1 \neq a_j \in H_j$ and p be the number of the a_j which are proper powers in H_j ($1 \leq j \leq n$). Let
$\{x_1, \cdots, x_m\}$ ($m \geq 1$) be any system in G and X be the subgroup of G generated by $x_1, \cdots, x_m$. If
$a_1 \cdots a_n \in X$ then one of the following cases occurs:*

(a) *$\{x_1, \cdots, x_m\}$ is Nielsen equivalent to a system $\{y_1, \cdots, y_m\}$ with $y_1 = a_1 \cdots a_n$.*

(b) *We have $m \geq 2n - p$, and $\{x_1, \cdots, x_m\}$ is Nielsen equivalent to a system $\{y_1, \cdots, y_m\}$ with
$y_i \in H_j$ ($1 \leq j \leq n$, $1 \leq i \leq 2n - p$).*   □

In [15] the interpretation of [2] in Lemma 2 of [1] was used for the proof of the application
Theorem (3.5) (see Lemma (3.1) of [15]). The above example shows that Lemma (3.1) of [15] is not
correct as stated. But, in fact, this lemma is not essential for the proof of Theorem (3.5) in [15]
under the restrictive pre-suppositions for this theorem. If we replace Lemma (3.1) of [14] by the
following theorem then the proof of Theorem (3.5) in [15] goes through as stated.

Theorem (2.12): *Let F be a free group with a free generating system B. Let $B_1, \cdots, B_n$ ($n \geq 1$)
be pairwise disjointed, nonempty subsets of B; let F_j be the subgroup of F freely generated by B_j
($1 \leq j \leq n$). Let $1 \neq w_j \in F_j$ and $w = w_1^{\alpha_1} \cdots w_n^{\alpha_n} \in F$, $\alpha_i \geq 1$ for $i = 1, \cdots, n$. Let $\{x_1, \cdots, x_m\}$
($1 \leq m \leq n$) be any system in F and X be the subgroup of F generated by $x_1, \cdots, x_m$. If $w^\alpha \in X$
for some $\alpha \neq 0$ then one of the following cases occurs:*

(a) *$\{x_1, \cdots, x_m\}$ is Nielsen equivalent to a system $\{y_1, \cdots, y_m\}$ with $y_1 = w^\beta$, $\beta \geq 1$, and β is
the smallest positive number for which a relation $w^\beta \in X$ holds.*

(b) *We have $m = n$, and $\{x_1, \cdots, x_m\}$ is Nielsen equivalent to a system $\{y_1, \cdots, y_m\}$ with*
$y_i = w_i^{\gamma_i}$, $1 < \gamma_i < \alpha_i$, $\gamma_i | \alpha_i$, ($i = 1, \cdots, m$).

<u>Proof.</u> We write F as the free product $F = F_1 * \cdots * F_n$ of the groups $F_1, \cdots, F_n$. Suppose
that a length L and a suitable order are introduced as in [4], [17] and [20]. We may assume that
$x_i \neq 1$ for all i and that $\{x_1, \cdots, x_m\}$ is Nielsen-reduced with respect to the length L and the order
(see [17]). The cases (1.3), (1.4) and (1.5) of Lemma (1.1) cannot hold. Hence, case (1.2) holds for
$\{x_1, \cdots, x_m\}$. In the following let α be the smallest positive number for which a relation $w^\alpha \in X$
holds. Analogously as in the proof of Theorem (2.9) we obtain that $\alpha = 1$ or case (a) holds; we
remark that we do not need a lemma like Lemma (2.3) which was crucial for the proof of Theorem
(2.9). If $\alpha = 1$ then we can apply Theorem (2.11) and obtain that case (a) holds or that $m = n$,
$\alpha_i \geq 2$ for $i = 1, \cdots, n$ and $\{x_1, \cdots, x_m\}$ is Nielsen equivalent to a system $\{y_1, \cdots, y_m\}$ with $y_i \in F_i$
for $i = 1, \cdots, n$ (we remark that, necessarily, if each x_i is contained in some F_j then no two of the
x_i are contained in the same F_j). This gives the result.   □

3. Minimal generating systems for plane discontinuous groups

In the following let

$$G = \langle s_1, \cdots, s_m, a_1, \cdots, a_p \mid s_1^{\gamma_1} = \cdots = s_m^{\gamma_m} = s_1 s_2 \cdots s_m P(a_1, \cdots, a_p) = 1\rangle$$

with $m \geq 0$, $p \geq 0$, all $\gamma_i \geq 2$ and

$$P(a_1, \cdots, a_p) = a_1^{\alpha_1} \cdots a_n^{\alpha_n} [a_{n+1}, a_{n+2}] \cdots [a_{p-1}, a_p]$$

$0 \leq n \leq p$, $p - n$ even, all $\alpha_j \geq 2$ such that

(i) $p \geq 2$ or
(ii) $p = 1$ and $m \geq 2$ or
(iii) $p = 0$ and $m \geq 4$ or
(iii) $p = 0$, $m = 3$ and $\frac{1}{\gamma_1} + \frac{1}{\gamma_2} + \frac{1}{\gamma_3} \leq 1$.

If $n = 0$ or $n = p$ and all $\alpha_i = 2$ then G is a plane discontinuous group. With the help of Theorem (2.2), Theorem (2.9) and the results of [8] we easily get the following.

Theorem (3.1) (see [8]): *The rank of G is*

(a) p if $m = 0$,
(b) $m - 2$ if $p = 0$, m is even, and all γ_i equal 2 except for one, which is odd,
(c) $p + m - 1$ in all other cases.   □

We are interested to describe the Nielsen equivalence classes of minimal generating systems for G. If $m \leq 1$ then the result is well-known.

Theorem (3.2) (see [11], [20], [21]): *Let $m \leq 1$ and $\{x_1, \cdots, x_p\}$ be a (minimal) generating system for G.*

(a) If $m = 1$ then $\{x_1, \cdots, x_p\}$ is Nielsen equivalent to $\{a_1, \cdots, a_p\}$.

(b) If $m = 0$ and $p \geq 3$ then $\{x_1, \cdots, x_p\}$ is Nielsen equivalent to a system $\{a_1, \cdots, a_{i-1}, a_i^{\beta_i}, a_{i+1}, \cdots, a_n, a_{n+1}, \cdots, a_p\}$ for some i $(1 \leq i \leq n)$ with (β_i, α_i) and $1 \leq \beta_i \leq \frac{1}{2}\alpha_i$.

(c) If $m = n = 0$ and $p = 2$ then $\{x_1, x_2\}$ is Nielsen equivalent to $\{a_1, a_2\}$.

(d) If $m = 0$ and $n = p = 2$ then $\{x_1, x_2\}$ is Nielsen equivalent to a system $\{a_1^{\beta_1}, a_2^{\beta_2}\}$ with $(\beta_1, \beta_2) = (\beta_1, \alpha_1) = (\beta_2, \alpha_2) = 1$ and $1 \leq \beta_1 \leq \frac{1}{2}\beta_2\alpha_1$, $1 \leq \beta_2 \leq \frac{1}{2}\beta_1\alpha_1$.   □

Remark: Zieschang [21] showed that in (3.2.d) $\{x_1, x_2\}$ is Nielsen equivalent to just one system $\{a_1^{\beta_1}, a_2^{\beta_2}\}$ with β_1, β_2 as above. The analogous statement holds in (3.2.b).

Theorem (3.3) (see [16] and the remarks in §2): *Let $p \geq 2$, $m \geq 2$ and $\{x_1, \cdots, x_{p+m-1}\}$ be a (minimal) generating system for G. Then $\{x_1, \cdots, x_{p+m-1}\}$ is Nielsen equivalent to a system $\{s_{\nu_1}^{\beta_1}, \cdots s_{\nu_{m-1}}^{\beta_{m-1}}, a_1, \cdots, a_p\}$ with $\nu_i \in \{1, \cdots, m\}$, $\nu_1 < \cdots < \nu_{m-1}$ (if $m \geq 3$) and $1 \leq \beta_i < \gamma_{\nu_i}$, $(\beta_i, \gamma_{\nu_i}) = 1$.*   □

Remark: Of course, we can get in addition $1 \leq \beta_i \leq \frac{1}{2}\gamma_{\nu_i}$, $(\beta_i, \gamma_{\nu_i}) = 1$.

Theorem (3.4): *Let $p = 1$, $m \geq 2$ and $\{x_1, \cdots, x_m\}$ be a (minimal) generating system for G. Suppose that one of the following holds:*

    (a) at least two γ_i are greater than 2,

    (b) all γ_i are equal to 2 except for one which is even and greater than 2.

    (c) m is odd, all γ_i are equal to 2 except for one which is odd and α_1 is even.

    (d) m is even and all γ_i are equal to 2.

Then $\{x_1, \cdots, x_m\}$ is Nielsen equivalent to a system $\{s_{\nu_1}^{\beta_1}, \cdots s_{\nu_{m-1}}^{\beta_{m-1}}, a_1\}$ with $\nu_i \in \{1, \cdots, m\}$, $\nu_1 < \cdots < \nu_{m-1}$ (if $m \geq 3$) and $1 \leq \beta_i < \gamma_{\nu_i}$, $(\beta_i, \gamma_{\nu_i}) = 1$.

Proof. We write a instead of a_1, α instead of α_1 and G as an amalgamated free product $G = H_1 *_A H_2$ with $H_1 = \langle s_1, \cdots, s_m \mid s_1^{\gamma_1} = \cdots = s_m^{\gamma_m} = 1 \rangle$, $H_2 = \langle a \rangle$ and $A = \langle a^\alpha \rangle = \langle s_1 \cdots s_m \rangle$. We assume that in G a length L and a suitable order are introduced as in [4], [17] and [20]. Without any loss of generality let $\{x_1, \cdots, x_m\}$ be Nielsen-reduced with respect to L and the order. We apply Lemma (1.1), Theorem (3.1) and Theorem (1.2) of [16] (see also Theorem 1 of [8]).

Suppose $x_i = z s_j^{\delta_j} z^{-1}$ $(1 \leq i, j \leq m)$ with $1 \leq \delta_j < \gamma_j$, $(\delta_j, \gamma_j) = d_j \geq 2$ and $z \in G$, and, without any loss of generality, let $j = 1$. Then the factor group

$$\bar{G} = \langle s_1, \cdots, s_m, a \mid s_1^{d_1} = s_2^{\gamma_2} = \cdots = s_m^{\gamma_m} = s_1 s_2 \cdots s_m a^\alpha = 1 \rangle$$

can be generated by the images $\bar{x}_1, \cdots, \bar{x}_{i-1}, \bar{x}_{i+1}, \cdots \bar{x}_m$ of $x_1, \cdots, x_{i-1}, x_{i+1}, \cdots, x_m$, that means, by $m-1$ elements. This contradicts Theorem (3.1). Hence, if x_i is conjugate to $s_j^{\delta_j}$, $1 \leq \delta_j < \gamma_j$, then $(\delta_j, \gamma_j) = 1$. Of course, $x_1, \cdots, x_m$ cannot all be contained in a subgroup of G which is conjugate to H_1. Therefore, we may assume that $x_1 = a^\delta$, $\delta \geq 1$ because at least one $\gamma_i \geq 3$ if m is odd. Hence, let $x_1 = a^\delta$, $\delta \geq 1$.

We show that $\delta = 1$. Suppose $\delta \geq 2$. Let first $m = 2$. Then $\delta|\alpha$ and $(\gamma_1, \gamma_2) = 1$ because a factor group $\langle s_1, s_2, a \mid s_1^{\gamma_1} = s_2^{\gamma_2} = a^\beta = s_1 s_2 a^\alpha = 1 \rangle$ with $\beta \mid \alpha$ or $(\gamma_1, \gamma_2) \geq 2$ is not cyclic. Hence, $\gamma_1, \gamma_2 \geq 3$, $(\gamma_1, \gamma_2) = 1$ and $\delta|\alpha$. Without any loss of generality, we may assume that $x_2 = h_1 \cdots h_k a^\beta$, $0 \leq \beta < \delta$, whereby the h_j are coset representatives of A in the H_i different from 1, with adjacent terms in distinct factors and $h_1, h_k \in H_1$. Now, it is easy to see that x_1 and x_2 can generate G only if $k = 1$ and $1 \leq \beta < \delta$, that is, $x_2 = h_1 a^\beta$ with $h_1 \in H_1$, $1 \leq \beta < \delta$. Then, necessarily, H_1 is generated by $a^{-\alpha} = s_1 s_2$ and $x_2 a^{-\alpha} x_2^{-1} = h_1 s_1 s_2 h_1^{-1}$ which gives a contradiction because a free product $\langle s_1, s_2 \mid s_1^{\gamma_1} = s_2^{\gamma_2} = 1 \rangle$ can be generated by two elements $s_1 s_2$ and $h_1 s_1 s_2 h_1^{-1}$ only if $\gamma_1 = 2$ or $\gamma_2 = 2$ (see for instance [19]). Hence, $\delta = 1$ if $m = 2$. Now, let $m \geq 3$. Then $(\delta, \alpha) = 1$ because a factor group

$$\langle s_1, \cdots, s_m \mid s_1^{\gamma_1} = \cdots = s_m^{\gamma_m} = s_1 \cdots s_m = 1 \rangle \ * \ \langle a \mid a^d = 1 \rangle$$

with $2 \leq d$ and $d|\alpha$ has rank m (see Theorem (3.1)). Hence, $(\delta, \alpha) = 1$. The factor group

$$\langle s_1, \cdots, s_m, s \mid s_1^{\gamma_1} = \cdots = s_m^{\gamma_m} = s^\delta = s_1 \cdots s_m s = 1 \rangle$$

has rank m under the presuppositions (see Theorem (3.1)); we remark that $\delta \geq 3$ if α is even. This gives a contradiction. Hence, also, $\delta = 1$ if $m \geq 3$.

Therefore, $x_1 = a$ in each case. Now we get a Nielsen transformation from $\{x_1, \cdots, x_m\}$ to a system $\{s_{\nu_1}^{\beta_1}, \cdots s_{\nu_{m-1}}^{\beta_{m-1}}, a\}$ with $\nu_i \in \{1, \cdots, m\}$, $\nu_1 < \cdots < \nu_{m-1}$ (if $m \geq 3$) and $1 \leq \beta_i < \gamma_{\nu_i}$, $(\beta_i, \gamma_{\nu_i}) = 1$ as in the proof of Theorem (2.2) of [16].

Remark: Of course, Theorem (3.4) is not best possible, but on the other side it does not hold in general. This can be seen from the following examples.

(a) Let $G = \langle s_1, s_2, a \mid s_1^2 = s_2^3 = s_1 s_2 a^2 = 1 \rangle$. Let $x = a^2$, $y = s_1 a$ and $X = \langle x, y \rangle$. Because $yx^{-1}y^{-1} = s_1 a^{-2} s_1 = s_2 s_1$ and $s_2 s_1 x^{-1} = s_2 s_1 a^{-2} = s_2^2 = s_2^{-1}$ we obtain $s_2 \in X$. Hence, $s_1 \in X$ and $a \in X$, that is $X = G$ and $\{x, y\}$ is a generating pair of G. This pair $\{x, y\}$ is not Nielsen equivalent to $\{s_1, a\}$ or $\{s_2, a\}$. This can be seen as follows. We regard G as an amalgamated free product $G = H_1 *_A H_2$ with $H_1 = \langle s_1, s_2 \mid s_1^2 = s_2^3 = 1 \rangle$, $H_2 = \langle a; \rangle$ and $A = \langle a^2 \rangle = < s_1 s_2 >$. We assume that in G a length L and and order are introduced as in the proof of Theorem (3.4). Then $[y, x] = s_1 a^2 s_1 a^{-2} = s_1 s_2^{-1} s_1 s_2 = [s_1, s_2^{-1}]$, hence $L([y, x]) = 1$. But $[s_1, a]$ and $[s_2, a]$ both are cyclically reduced of length 4, that means, $[s_1, a]$, and also $[s_2, a]$, cannot be conjugate in G to $[y, x]^\varepsilon$, $\varepsilon = \pm 1$. Therefore, $\{x, y\}$ is not Nielsen equivalent to $\{s_1, a\}$ or $\{s_2, a\}$ (see [7, p.44]).

(b) Let $G = \langle s_1, \cdots, s_m, a \mid s_1^2 = \cdots = s_m^2 = s_1 \cdots s_m a^5 = 1 \rangle$ with $m = 2k + 1$, $k \geq 1$. Let $x_1 = a^3$, $x_i = s_1 s_i$ for $i = 2, \cdots, m$ and $X = \langle x_1, \cdots, x_m \rangle$. Then $(s_1 \cdots s_m)^{-2} = a^{10} \in X$ (see Theorem 1 of [8]). Hence, $a \in X$, $a^{-5} = s_1 \cdots s_m \in X$ and $s_1, \cdots, s_m \in X$, that is, $X = G$.

(c) Let $G = \langle s_1, s_2, s_3, a \mid s_1^2 = s_2^2 = s_3^7 == s_1 s_2 s_3 a^5 = 1 \rangle$. Let Let $x_1 = a^2$, $x_2 = a s_1$, $x_3 = a s_2$ and $X = \langle x_1, x_2, x_3 \rangle$. Then $x_1^2 x_2 x_3^{-1} x_1^3 x_2^{-1} x_3 = (a^5 s_1 s_2)^2 = s_3^{-2}$. Hence, $s_3 \in X$, $x_1^{-2}(a^5 s_1 s_2) x_3^{-1} x_2 = a \in X$, $s_1 \in X$, $s_2 \in X$, that is, $X = G$.

From now on, let $p = 0$, that is

$$G = \langle s_1, \cdots, s_m \mid s_1^{\gamma_1} = \cdots = s_m^{\gamma_m} = s_1 \cdots s_m = 1 \rangle$$

with $m \geq 3$ and $\frac{1}{\gamma_1} + \frac{1}{\gamma_2} + \frac{1}{\gamma_3} \leq 1$ if $m = 3$.

The case (b) of Theorem (3.1), where rank $G = m - 2$, is particularly interesting. Here there is just one Nielsen equivalence class of generating systems $\{x_1, \cdots, x_{m-2}\}$, more concrete, if m is even, $\gamma_1 = \cdots = \gamma_{m-1} = 2$ and γ_m odd then each generating system $\{x_1, \cdots, x_{m-2}\}$ is Nielsen equivalent to $\{s_1 s_2, \cdots, s_1 s_{m-1}$ (see Theorem (2.12) of [16]). If $m = 3$ and $\frac{1}{\gamma_1} + \frac{1}{\gamma_2} + \frac{1}{\gamma_3} = 1$ then, – up to permutations – $(\gamma_1, \gamma_2, \gamma_3) = (3,3,3)$, $(2,4,4)$ or $(2,3,6)$, and G is an Euclidean, plane discontinuous group and, hence, the generating pairs for G are well-known. If $m = 3$ and $\frac{1}{\gamma_1} + \frac{1}{\gamma_2} + \frac{1}{\gamma_3} < 1$ the generating pairs for G are completely described in [19]. Especially, if $m = 3$ and $\gamma_i \geq 4$ for $i = 1, 2, 3$, then any generating pair of G is Nielsen equivalent to a pair $\{s_{\nu_1}^{\beta_1}, s_{\nu_2}^{\beta_2}\}$ with $\nu_i \in \{1, 2, 3\}$, $\nu_1 < \nu_2$, $1 \leq \beta_i < \gamma_{\nu_i}$, $(\beta_i, \gamma_{\nu_i}) = 1$ $(i = 1, 2)$.

Theorem (3.5): *Let* $p = 0$, $m \geq 4$, *all* $\gamma_i \geq 4$ *and* $\{x_1, \cdots, x_{m-1}\}$ *be a (minimal) generating system for* G. *Then* $\{x_1, \cdots, x_{m-1}\}$ *is Nielsen equivalent to a system* $\{s_{\nu_1}^{\beta_1}, \cdots s_{\nu_{m-1}}^{\beta_{m-1}}\}$ *with* $\nu_i \in \{1, \cdots, m\}$, $\nu_1 < \cdots < \nu_{m-1}$ *and* $1 \leq \beta_i < \gamma_{\nu_i}$, $(\beta_i, \gamma_{\nu_i}) = 1$.

Remark: The proof of Theorem (3.5) is rather technical and complicated. If $m \geq 6$ then a proof also can be found in [9]. Hence, we here prove Theorem (3.5) only for $m = 4$ and $m = 5$.

Proof of Theorem (3.5) for $m = 4$ and $m = 5$.

Let $m = 4$ or $m = 5$. We write G as an amalgamated free product $G = H_1 *_A H_2$ with $H_1 = \langle s_1, s_2 \mid s_1^{\gamma_1} = s_2^{\gamma_2} = 1 \rangle$, $H_2 = \langle s_3, \cdots, s_m \mid s_3^{\gamma_3} = \cdots = s_m^{\gamma_m} = 1 \rangle$, and $A = \langle s_1 s_2 \rangle = \langle s_3 \cdots s_m \rangle$.

We assume that in G a length L and a suitable order are introduced as in [4], [17] and [20]. Without any loss of generality let $\{x_1, \cdots, x_m\}$ be Nielsen reduced with respect to L and the order. We apply Lemma (1.1), Theorem (3.1), Theorem (1.2) of [16] and Theorems 1 and 2 of [18] and Lemma 4 of [20]. Analogously as in the proof of Theorem (3.4) we get that if $x_i = z s_j^{\delta_j} z^{-1}$ ($1 \leq i \leq m-1$, $1 \leq j \leq m$) with $1 \leq \delta_j < \gamma_j$ and $z \in G$ then $(\delta_j, \gamma_j) = 1$. Also, if x_i is conjugate to a power of some s_{ν_i} and x_j is conjugate to a power of some s_{ν_j} ($1 \leq i, j \leq m-1$, $i \neq j$, $1 \leq \nu_i, \nu_j \leq m$) then $\nu_i \neq \nu_j$. This follows easily from Theorem (3.1). In the following we make use of these two facts. Also, we always have in mind that all $\gamma_i \geq 4$. Case (1.3) and (1.5) of Lemma (1.1) cannot occur (see Lemma 4 of [20]). If case (1.2) of Lemma (1.1) holds then, necessarily, $L(x_i) \leq 1$ for $i = 1, \cdots, m$, that is, each x_i is contained in H_1 or H_2. Of course just one of the x_i is contained in A, just one of the x_i is contained in $H_1 \setminus A$ and just $m-3$ of the x_i are contained in $H_2 \setminus A$, say $x_1 \in H_1 \setminus A$, $x_2 \in A$ and $x_3, \cdots, x_{m-1} \in H_2 \setminus A$. Then, necessarily, $H_1 = \langle x_1, x_2 \rangle$ and $H_2 = \langle x_2, \cdots, x_{m-1} \rangle$ which easily gives the desired result. This proves Theorem (3.5) if case (1.2) of Lemma (1.1) holds. For the following we remark that, clearly, if q ($q \geq 1$) of the x_i are contained in H_j ($j = 1$ or 2) then $q \leq 2$ if $j = 1$ and $q \leq 3$ if $j = 2$.

(3.6) Let $m = 4$.

Without any loss of generality, we may assume that x_1 and x_2 are contained in H_1 and a certain product of them is conjugate to a non-trivial power of $s_1 s_2$ because of Lemma (1.1) (and the succeeding remark) and the above remarks. The conjugacy factor can be taken to be trivial. If we apply the techniques of [8] we see that we may assume that one of the following cases occurs:

(1) $\{x_1, x_2\}$ is Nielsen equivalent to a pair $\{y_1, y_2\}$ with $y_1 = (s_1 s_2)^\delta$, $\delta \geq 1$.

(2) $\{x_1, x_2\}$ is Nielsen equivalent to a pair $\{y_1, y_2\}$ with $y_1 = s_1^{\beta_1}$, $1 \leq \beta_1 \leq \gamma_1$.

Case (1). $\{x_1, x_2\}$ is Nielsen equivalent to a pair $\{y_1, y_2\}$ with $y_1 = (s_1 s_2)^\delta$, $\delta \geq 1$. Then $\delta = 1$ because a group $\langle s_3, s_4 \mid s_3^{\gamma_3} = s_4^{\gamma_4} = (s_3 s_4)^\gamma = 1 \rangle$ with $\gamma \geq 2$ is non-cyclic. If we apply the Nielsen reduction method to $\{y_1, y_2, x_3\}$ then we, necessarily, get that the result holds or that $\{y_1, y_2\}$ is Nielsen equivalent to a pair $\{s_1^{\beta_1}, s_2^{\beta_2}\}$, $1 \leq \beta_i < \gamma_i$, $(\beta_i, \gamma_i) = 1$ because it cannot be that $x_1 = (s_1 s_2)^{\rho_1}$, $x_2 = g(s_1 s_2)^{\rho_2} g^{-1}$ and $x_1^\varepsilon x_2^\eta = h(s_1 s_2)^{\rho_3} h^{-1}$ with all $\rho_i \in \mathbf{Z} \setminus \{0\}$, $g, h \in H_1$ and $\varepsilon, \eta = \pm 1$ (we recall that $\gamma_i \geq 4$ for all i). We regard the latter. We may assume that the normal form of x_3 neither starts nor ends with a factor from H_1. Then, necessarily, $L(x_3) \leq 1$ which easily gives the desired result.

Case (2). $\{x_1, x_2\}$ is Nielsen equivalent to a pair $\{y_1, y_2\}$ with $y_1 = s_1^{\beta_1}$, $1 \leq \beta_1 < \gamma_1$. Then $(\beta_1, \gamma_1) = 1$. Hence, we may use the techniques in [8] correspondingly (see also the proof of Theorem (2.9)) and get that $\{y_1, y_2\}$ is Nielsen equivalent to a system $\{s_1^{\beta_1}, (s_1 s_2)^\delta\}$, $\delta \geq 1$, or to a system $\{s_1^{\beta_1}, s_2^{\beta_2}\}$, $1 \leq \beta_2 < \gamma_2$, $(\beta_2, \gamma_2) = 1$. In both cases we get the desired result (see the above case (1)). This proves the result for $m = 4$.

(3.7) Let $m = 5$.

We remark that a group $\langle s_3, s_4, s_5 \mid s_3^{\gamma_3} = s_4^{\gamma_4} = s_5^{\gamma_5} = (s_3 s_4 s_5)^\gamma = 1 \rangle$ with $\gamma \geq 2$ has rank 3. Therefore, with analogous arguments as in (3.6) the result follows easily if two of the x_i are contained in a subgroup which is conjugate to H_1 and a product of them is conjugate to a non-trivial power of $s_1 s_2$. Correspondingly, the result easily follows if three of the x_i are contained in a subgroup which is conjugate to H_2 and a certain product of them is conjugate to a non-trivial power of $s_3s_4s_5$ (we remark that 3 is odd and use the techniques of [8] and, especially, Lemma (2.1) of [8]).

Hence, without any loss of generality, we now may assume that x_3 and x_4 are contained in H_2 and a certain product of them is conjugate to a non-trivial power of $s_3s_4s_5$ because of Lemma (1.1). The conjugacy factor can be taken to be trivial. If we apply the techniques of [8] we see that we may assume that one of the following cases occurs:

(1) $\{x_3, x_4\}$ is Nielsen equivalent to a pair $\{y_3, y_4\}$ with $y_4 = (s_3s_4s_5)^\delta$, $\delta \geq 1$.

(2) $\{x_3, x_4\}$ is Nielsen equivalent to a pair $\{y_3, y_4\}$ with $y_4 = s_5^{\beta_5}$, $1 \leq \beta_5 < \gamma_5$.

<u>Case (1)</u>. $\{x_3, x_4\}$ is Nielsen equivalent to a pair $\{y_3, y_4\}$ with $y_4 = (s_3s_4s_5)^\delta$, $\delta \geq 1$. We write x_3, x_4 instead of y_3, y_4 (possibly after renaming). We show that $\delta = 1$. Suppose $\delta \geq 2$. Without any loss of generality, we may assume that δ is prime. We introduce the relation $(s_3s_4s_5)^\delta = 1$ and get the factor group $\bar{G} = K_1 *_B K_2$ with $K_1 = \langle s_1, s_2 \mid s_1^{\gamma_1} = s_2^{\gamma_2} = (s_1s_2)^\delta = 1 \rangle$, $K_2 = \langle s_3, s_4, s_5 \mid s_3^{\gamma_3} = s_4^{\gamma_4} = s_5^{\gamma_5} = (s_3s_4s_5)^\delta = 1 \rangle$ and $B = \langle s_1s_2 \mid (s_1s_2)^\delta = 1 \rangle = \langle s_3s_4s_5 \mid (s_3s_4s_5)^\delta = 1 \rangle$. We introduce in $\bar{G}$ a length ℓ and a suitable order as in [4], [17] and [20]. We regard the generating system $\{x_1, x_2, x_3\}$ of $\bar{G}$. Since $x_3 \in K_2$, $\ell(x_3) \leq 1$, we may assume that $\{x_1, x_2, x_3\}$ is Nielsen reduced with respect to the length ℓ and the order (if $\ell(x_3) = 1$ then possibly after replacing x_3 by an element from B which we denote again by x_3). Case (1.3) and case (1.5) of Lemma (1.1) cannot occur. This is trivial if $\gamma_1 = \gamma_2$ because then the factor group $\langle s_1 \mid s_1^{\gamma_1} = 1 \rangle * \langle s_3, s_4, s_5 \mid s_3^{\gamma_3} = s_4^{\gamma_4} = s_5^{\gamma_5} = s_3s_4s_5 = 1 \rangle$ has rank 3. If $\gamma_1 \neq \gamma_2$ we may use the fact that K_1 and K_2 are plane discontinuous subgroups of $PSL(2,\mathbb{R})$ and, hence, if $x(s_1s_2)^{\rho_2}x^{-1} = (s_1s_2)^{\rho_2}$, $x \in K_1$, $\rho_1, \rho_2 \in \mathbb{Z} \setminus \{0\}$, in K_1 then x is a power of s_1s_2; and analogously for K_2. x_1 and x_2 cannot both be contained in a subgroup which is conjugate to K_1 because the factor group $\langle s_3, s_4, s_5 \mid s_3^{\gamma_3} = s_4^{\gamma_4} = s_5^{\gamma_5} = s_3s_4s_5 = 1 \rangle$ is non-cyclic. They also cannot both be contained in K_2 because already $x_3 \in K_2$. Let $x_2 \in K_2$ and a product in x_2 and x_3 be conjugate to a non-trivial power of $s_3s_4s_5$ (the conjugacy factor can here be taken to be trivial). Then $\langle x_2, x_3 \rangle$ has infinite index in K_2, and, hence, $\langle x_2, x_3 \rangle$ is a free product of two cyclic groups because, clearly, it is not cyclic (we recall that all $\gamma_i \geq 4$). Without any loss of generality, we, therefore, may assume that $x_3 = s_3s_4s_5$ (possibly after renaming, we recall that δ is prime). If a free product $\langle a, b \mid a^\delta = b^q = 1 \rangle$, $q = 0$ or $q \geq 2$, is generated by a and d where the normal form (with respect to this free product) of d neither starts nor ends with a power of a then no cyclically reduced word in a and d which contains d is conjugate to a non-trivial power of a.

Hence, we may assume that case (1.2) of Lemma (1.1) holds if x_2 and x_3 are conjugate to a non-trivial power of $s_3s_4s_5$. The analogous statement (up to a conjugacy factor) holds if x_1 and x_2 both are contained in a subgroup which is conjugate to K_2 and a product in them is conjugate to a non-trivial power of $s_3s_4s_5$. Altogether, we may assume that case (1.2) of Lemma (1.1) holds. Then $\ell(x_i) \leq 1$ for $i = 1, 2, 3$ and, necessarily, $\ell(x_i) = 0$ for some i ($1 \leq i \leq 3$). Hence, K_1 is cyclic or K_2 has rank 2 which gives a contradiction. Therefore, $\delta = 1$, that is, $x_4 = s_3s_4s_5$.

Now, Lemma (1.1), the techniques in [8] and the arguments until now show that the result holds or that we may assume that $\{x_3, x_4\}$ is Nielsen equivalent to a pair $\{z_3, z_4\}$ with $z_4 = z_i^{\beta_i}$, $1 \leq \beta_i < \gamma_i$, for some i, $3 \leq i \leq 5$ (change the decomposition as an amalgamated free product if convenient). Without any loss of generality, in the latter we may assume $z_4 = s_5^{\beta_5}$. Hence, we can reduce case (1) to case (2).

<u>Case (2)</u>. $\{x_3, x_4\}$ is Nielsen equivalent to a pair $\{y_3, y_4\}$ with $y_4 = s_5^{\beta_5}$, $1 \leq \beta_5 < \gamma_5$. By the above arguments, $(\beta_5, \gamma_5) = 1$. We write x_3, x_4 instead of y_3, y_4 (possibly after renaming). Now we regard

another decomposition $G = N_1 *_C N_2$ as an amalgamated free product with $N_1 = \langle s_5, s_1 \mid s_5^{\gamma_5} = s_1^{\gamma_1} = 1 \rangle$, $N_2 = \langle s_2, s_3, s_4 \mid s_2^{\gamma_2} = s_3^{\gamma_3} = s_4^{\gamma_4} = 1 \rangle$ and $C = \langle s_5 s_1 \rangle = \langle s_2 s_3 s_4 \rangle$.

Now, let the length L and the order be referred to this decomposition. We reduce $\{x_1, x_2, x_3, x_4\}$ with respect to the length and the order. Since $x_4 \in N_1$, $L(x_4) = 1$, we may assume that $\{x_1, x_2, x_3, x_4\}$ is Nielsen reduced or that x_3 is contained in N_1 and a product in x_3 and x_4 is conjugate to a non-trivial power of $s_5 s_1$. In the latter we may possibly replace x_4 by a power of $s_5 s_1$ to get a Nielsen reduced system. But the above arguments show that in the latter the result holds. Now, let $\{x_1, x_2, x_3, x_4\}$ be Nielsen reduced. Then the above arguments show that the result holds or $\{x_1, x_2, x_3, x_4\}$ is Nielsen equivalent to a system $\{y_1, y_2, y_3, y_4\}$ with $y_1 = s_5^{\beta_5}$ and $y_2 = z s_i^{\beta_i} z^{-1}$, $z \in G$, $1 \le \beta_i < \gamma_i$, $(\beta_i, \gamma_i) = 1$ for some i ($2 \le i \le 4$).

We regard the latter. Without any loss of generality, let $y_1 = s_5^{\beta_5}$ and $y_2 = z s_4^{\beta_4} z^{-1}$. Now we consider a third decomposition $G = M_1 *_D M_2$ as an amalgamated free product with $M_1 = \langle s_4, s_5 \mid s_4^{\gamma_4} = s_5^{\gamma_5} = 1 \rangle$, $M_2 = \langle s_1, s_2, s_3 \mid s_1^{\gamma_1} = s_2^{\gamma_2} = s_3^{\gamma_3} = 1 \rangle$ and $D = \langle s_4 s_5 \rangle = \langle s_1 s_2 s_3 \rangle$. Let the length L and the order be referred to this decomposition. We reduce $\{y_1, y_2, y_3, y_4\}$ with respect to the length and the order and use, in addition, the techniques of [4]. If $y_1^\varepsilon y_2^\eta$ is conjugate to a non-trivial power of $s_4 s_5$ for some $\varepsilon, \eta = \pm 1$ then, necessarily, $y_2 \in M_1$ and the result holds by the above arguments. Hence, we now assume that no product $y_1^\varepsilon y_2^\eta$, $\varepsilon, \eta = \pm 1$, is conjugate to a non-trivial power of $s_4 s_5$. Of course, no product, $y_i^\varepsilon y_2^\eta$, $\varepsilon, \eta = \pm 1$, $i = 3$ or 4, is conjugate to a non-trivial power of $s_4 s_5$.

If $L(y_i) > L(y_i^\varepsilon y_2^\eta)$ for some $i \in \{3, 4\}$, $\varepsilon, \eta = \pm 1$, then we replace y_i by $y_i^\varepsilon y_2^\eta$. If $L(y_i) \le L(y_i^\varepsilon y_2^\eta) < L(y_2)$ for some $i \in \{1, 3, 4\}$, $\varepsilon, \eta = \pm 1$, then also $L(y_i^\varepsilon y_2^\eta y_i^{-\varepsilon}) < L(y_2)$ because no such product $y_i^\varepsilon y_2^\eta$ is conjugate to a non-trivial power of $s_4 s_5$, and in this case we replace y_2 by $y_i^\varepsilon y_2^\eta y_i^{-\varepsilon}$. If $L(y_i^\varepsilon y_2^\eta) \ge L(y_i)$, $L(y_2)$ and $y_i^\varepsilon y_2^\eta$ stands before y_2 for some $i \in \{1, 3, 4\}$, $\varepsilon, \eta = \pm 1$, then $y_i, y_2 \in M_1$ or $L(y_i) = L(y_2) = L(y_i^\varepsilon y_2^\eta)$ and $y_i^\varepsilon y_2^\eta y_i^{-\varepsilon}$ stands before y_2, and in the latter we replace y_2 by $y_i^\varepsilon y_2^\eta y_i^{-\varepsilon}$.

Hence, by the above argument, we may assume that the result holds or that $\{y_1, y_2, y_3, y_4\}$ is Nielsen equivalent to a system $\{z_1, z_2, z_3, z_4\}$ with $z_1 = s_5^{\beta_5}$ and $z_2 = g s_4^{\beta_4} g^{-1}$, $g \in G$, and $z_3 = h s_i^{\beta_i} h^{-1}$, $h \in G$, $1 \le \beta_i < \gamma_i$, $(\beta_i, \gamma_i) = 1$ for some i ($1 \le i \le 3$). We regard the latter. Without any loss of generality, let $z_3 = h s_3^{\beta_3} h^{-1}$. Now, we again consider the first decomposition $G = H_1 *_A H_2$ and refer the length L and the order again to this decomposition. We reduce $\{z_1, z_2, z_3, z_4\}$ with respect to the length and the order. The above arguments, analogously used, show that, without any loss of generality, we may assume that $g, h \in H_2$. This, easily, gives the desired result. $\square$

Remarks:

1) Of course, Theorem (3.5) is not best possible, for instance, on can show that it also holds if we replace "all $\gamma_i \ge 4$" by "all $\gamma_i \ge 3$". The proof for this is more technical and complicated, it includes still more different cases. If $m \ge 6$ then a proof for this more general result can also be found in [9]. Possibly we can extend Theorem (3.5) so much that we may replace "all $\gamma_i \ge 4$" by "at least three $\gamma_i \ge 3$". But the general case seems to be very difficult. This can be seen from the following example: Let $G = \langle s_1, s_2, s_3, s_4 \mid s_1^{10} = s_2^2 = s_3^2 = s_4^5 = s_1 s_2 s_3 s_4 = 1 \rangle$. Let $x_1 = s_1^4$, $x_2 = s_1 s_2$, $x_3 = s_1 s_3$ and $X = \langle x_1, x_2, x_3 \rangle$. Then $x_2 x_3^{-1} x_1^{-2} x_2^{-1} x_3 = (s_1 s_2 s_3)^2 = s_4^{-2} \in X$, hence, $s_4, s_1 s_2 s_3, s_3, s_1, s_2 \in X$, that is $X = G$.

2) In [5] Lustig and Moriah describe conditions under which two minimal generating systems of G are not Nielsen equivalent. They use these conditions to investigate the Heegard splittings of Seifert fibered 3-manifolds (see also [3]). In [6] we consider automorphisms of Fuchsian groups and their lifts to automorphisms of free groups with respect to minimal generating systems.

References

[1] B. Baumslag and F. Levin, *A free product with a non-power amalgamated which is not residually free,* Math Z., 151(1976), 235-237.

[2] G. Baumslag, *Residual nilpotence and relations in free groups,* J. Algebra 2 (1965), 271-282.

[3] M. Boileau and H. Zieschang, *Heegard genus of closed orientable Seifert manifolds,* Invent. Math. 76 (1984), 455-468.

[4] R.N. Kalia and G. Rosenberger, *Über Untergruppen ebener diskontinuierlicher Gruppen,* Contemporary Math. 33(1984), 308-327.

[5] M. Lustig and Y. Moriah, *Nielsen equivalence in Fuchsian groups and Seifert fibered spaces,* preprint 1987.

[6] M. Lustig, Y. Moriah and G. Rosenberger, *Automorphisms of Fuchsian groups and their lifts to free groups,* preprint 1987.

[7] R.C. Lyndon and P.E. Schupp, *Combinatorial group theory,* Ergebnisse der Math. 189, Springer, Berlin-Heidelberg-New York, 1977.

[8] N. Peczynski, G. Rosenberger and H. Zieschang, *Über Erzeugende ebener diskontinuierlicher Gruppen,* Invent. Math., 29(1975), 161-180.

[9] N. Peczynski, *Über Erzeugendensysteme von Fuchsschen Gruppen,* Dissertation, Bochum 1975.

[10] G. Rosenberger, *Zum Rang- und Isomorphieproblem für freie Produkte mit Amalgam,* Habilitationsschrift, Hamburg 1974.

[11] G. Rosenberger, *Zum Isomorphieproblem für Gruppen mit einer definierenden Relation,* Ill. J. Math. 20(1976), 614-621.

[12] G. Rosenberger, *Produkte von Potenzen und Kommutatoren in freien Gruppen,* J. Algebra, 53(1978), 416-422.

[13] G. Rosenberger, *Alternierende Produkte in freien Gruppen,* Pacific J. Math. 78 (1978), 243-250.

[14] G. Rosenberger, *Applications of Nielsen's reduction method to the solution of combinatorial problems in group theory: a survey,* London Math. Soc. Lecture Notes Series 36 (1979), 339-358.

[15] G. Rosenberger, *Über alternierende Wörter in freien Produkten mit Amalgam*, Archiv. Math. 31(1978), 417-422.

[16] G. Rosenberger, *Automorphismen ebener diskontinuierlicher Gruppen*, Proc. of the Stony Brook Conf. on Riemann surfaces and related topics 1978; Annals of Math. Studies 97, 439-455, Princeton Univ. Press, 1981.

[17] G. Rosenberger, *Gleichungen in freien Produkten mit Amalgam*, Math. Z. 173(1980), 1-11; Berichtigung, Math. Z. 178(1981), 579.

[18] G. Rosenberger, *Über Darstellungen von Elementen und Untergruppen in freien Produkten*, Proc. of 'Groups–Korea 1983', Lecture Notes in Math. 1098, Springer, Berlin-Heidelberg-New York, 1984.

[19] G. Rosenberger, *All generating pairs of all two-generator Fuchsian groups*, Archiv Math., 46(1986), 198-204.

[20] H. Zieschang, *Über die Nielsensche Kürzungsmethode in freien Gruppen mit Amalgam*, Invent. Math., 10(1970), 4-37.

[21] H. Zieschang, *Generators of the free product with amalgamation of two infinite cyclic groups*, Math. Ann. 227(1977), 195-221.

[22] H. Zieschang, E. Vogt and H.-D. Coldewey, *Surfaces and planar discontinuous groups*, Lecture Notes in Math. 835, Springer, Berlin-Heidelberg-New York 1980.

Fachbereich Mathematik
Universität Dortmund
Postfach 50 05 00
4600 Dortmund 50
Federal Republic of Germany

APPENDIX A

Talks presented at "Groups – Korea 1988". An asterisk (*) marks those talks that are reproduced (in edited form) in these Proceedings.

(1) List of invited one-hour talks:

Bernhard Amberg, *Infinite factorized groups*
Hyman Bass, *Group actions on trees* (I, II, III)
Michel M. Deza, *Sharp groups*
Walter Feit, *On Galois groups* (I, II, III)
Narain Gupta, *Higher dimension subgroups*
Noboru Ito, *Doubly regular asymmetric digraphs with rank 5 automorphism groups*
Otto H. Kegel, *Some locally finite groups* (I, II)
Jens L. Mennicke, *On Fibonacci groups*
Horace Y. Mochizuki, *Automorphism groups of relatively free groups*
Peter M. Neumann, *Pathological representations of infinite soluble groups* (I, II, III)
Michael F. Newman, *Proving a group infinite*
Stephen J. Pride, *Some problems in combinatorial group theory*
Akbar H. Rhemtulla, *Groups with many elliptic subgroups*
Derek J.S. Robinson, *Reflection on the construction theory of polycyclic groups* (I, II)
Klaus W. Roggenkamp, *On group rings*
Gerhard Rosenberger, *Minimal generating systems for plane discontinuous groups and an equation in free groups*

(2) List of seminar talks:

F. Rudolf Beyl, *Efficient presentations for modular special linear groups*
Emilio Bujalance, *The groups of automorphisms of non-orientable hyperelliptic Klein surfaces without boundary*
Colin M. Campbell, *Efficient presentations for finite simple groups and related groups*
Jung Rae Cho, *On n-groupoids defined on fields*
Martyn Dixon, *Divisible automorphism groups*
Dietmar Garbe, *Non-orientable and orientable regular maps*
C. Kanta Gupta, *Automorphisms of certain relatively free soluble groups*
Bruno Kahn, *Stiefel-Whitney classes of finite group representations*
Naoki Kawamoto, *Extensions of the class of Lie algebras with the lattice of subideals*
In Sok Lee, *Shintani liftings of irreducible modular representations*
Patrizia Longobardi, *On groups with a permutational property on commutators*
Mohammad R.R. Moghaddam, *On generalized Schur multiplicators*
B.H. Neumann, *Laws in a class of groupoids*
Markku Niemenmma, *On multiplication groups of loops*
Jae Keol Park, *Jacobson group algebras*
Carlo M. Scoppola, *Groups of Frobenius type*
Richard M. Thomas, *On certain one-relator products of cyclic groups*

APPENDIX B

List of Participants

Dr Hassan A. Al-Zaid, Kuwait University, Kuwait

Professor Bernhard Amberg, Universität Mainz, Federal Republic of Germany

Mr Dae Hyeon Baik, Pusan National University, Korea

Professor Yong Bai Baik, Hyosung Women's University, Korea

Mr Young Gheel Baik, Pusan National Susan University, Korea

Professor Hyman Bass, Columbia University, U.S.A.

Professor F.R. Beyl, Portland State University, U.S.A.

Dr Emilio Bujalance, Facultad De Ciencias U.N.E.D., Spain

Dr Colin M. Campbell, University of St. Andrews, Scotland

Professor A.E. Caranti, Universita Delgi Studi di Trento, Italy

Professor Kun Soo Chang, Yonsei University, Korea

Mr Yeon Kwan Cheong, Seoul National University, Korea

Dr K. Chitti, Mashhad University, Iran

Dr Jung Rae Cho, Pusan National University, Korea

Professor Michel M. Deza, C.N.R.S. France

Dr Martyn Dixon, University of Alabama, U.S.A.

Professor Walter Feit, Yale University, U.S.A.

Dr Dietmar Garbe, Universität Bielefeld, Federal Republic of Germany

Professor C.K. Gupta, University of Manitoba, Canada

Professor N.D. Gupta, University of Manitoba, Canada

Professor Ki Sik Ha, Pusan National University, Korea

Dr Woo Chorl Hong, Pusan National University, Korea

Dr Chan Huh, Pusan National University, Korea

Professor Won Huh, Pusan National University, Korea

Miss Akiko Ichinohe, University of Tokyo, Japan

Dr Toshiharu Ikeda, Hiroshima University, Japan

Professor Noboru Ito, Konan University, Japan

Dr David Johnson, University of Nottingham, England

Dr Bruno Kahn, C.N.R.S., France

Dr Naoki Kawamoto, Hiroshima University, Japan

Professor Otto H. Kegel, Albert–Ludwigs–Universität, Federal Republic of Germany

Dr. Reyadh Khazal, Kuwait University, Kuwait

Professor U – Hang Ki, Kyungpook National University, Korea

Professor Ann Chi Kim, Pusan National University, Korea

Mr Boo Yoon Kim, Pusan National University, Korea

Professor Chol On Kim, Pusan National University, Korea

Mr Hong Gi Kim, Pusan National University, Korea

Dr Jai Heui Kim, Pusan National University, Korea

Miss Jeong Young Kim, Pusan National University, Korea
Dr Myung Hwan Kim, Korea Institute of Technology, Korea
Mr Pan Soo Kim, University of Alberta, Canada
Mr Yangkok Kim, Pusan National University, Korea
Dr Ki Hyoung Ko, Korea Institute of Technology, Korea
Dr Ja Kyung Koo, Korea Institute of Technology, Korea
Professor Chung Gul Lee, Pusan National University, Korea
Professor Hyung Kyi Lee, Pusan National University, Korea
Dr In Sok Lee, Seoul National University, Korea
Dr Patrizia Longobardi, University of Napoli, Italy
Dr Toru Maeda, Kansai University, Japan
Dr Ernesto Martinez, Facultad De Ciencias U.N.E.D., Spain
Dr Osamu Maruo, Hiroshima University, Japan
Professor Jens L. Mennicke, Universität Bielefeld, Federal Republic of Germany
Professor Horace Mochizuki, University of California – Santa Barbara, U.S.A.
Dr M.R.R. Moghaddam, Mashhad University, Iran
Professor B.H. Neumann, Australian National University, Australia
Dr Peter M. Neumann, The Queen's College, Oxford University, England
Dr M.F. Newman, Australian National University, Australia
Professor Markku Niemenmma, University of Oulu, Finland
Professor Jae Keol Park, Pusan National University, Korea
Dr Jong Yeoul Park, Pusan National University, Korea
Professor Young Sik Park, Pusan National University, Korea
Professor Young Soo Park, Kyungpook National University, Korea
Dr Stephen Pride, University of Glasgow, Scotland
Professor Akbar H. Rhemtulla, University of Alberta, Canada
Professor Jeong Dae Rim, Yonsei University, Korea
Professor Derek J.S. Robinson, University of Illinois – Urbana Champaign U.S.A.
Professor K.W. Roggenkamp, Universität Stuttgart, Federal Republic of Germany
Professor Gerhard Rosenberger, Universität Dortmund, Federal Republic of Germany
Dr C.M. Scoppola, Universita Delgi Studi di Trento, Italy
Professor Tae Yeong Seo, Pusan National University, Korea
Dr Hyun Yong Shin, Korea Teacher's College, Korea
Professor Adeleide Sneider, University of Rome, Italy
Dr Kwang Ho Sohn, Pusan National University, Korea
Dr Hyeon Jong Song, Pusan National Susan University, Korea
Professor Tai Sung Song, Pusan National University, Korea
Professor Ken–Ichi Tahara, Aichi University of Education, Japan
Dr R.M. Thomas, University of Leicester, England
Dr R.F. Turner–Smith, Hong Kong Polytechnic, Hong Kong
Professor Jang Il Um, Pusan National University, Korea
Professor K. Varadarajan, University of Calgary, Canada
Professor Zhexian Wan, Academia Sinica, Bejing, People's Republic of China